Other Books by Richard Rhodes

FICTION

Sons of Earth

The Last Safari

Holy Secrets

The Ungodly

VERITY

Trying to Get Some Dignity (with Ginger Rhodes)

Dark Sun

How to Write

Nuclear Renewal

Making Love

A Hole in the World

Farm

The Making of the Atomic Bomb

Looking for America

The Inland Ground

Classic medical detective story."

—George Johnson,

The New York Times Book Review

DEADLY FEASTS

The "Prion" Controversy and the Public's Health

RICHARD RHODES

A Touchstone Book
Published by Simon & Schuster

TOUCHSTONE
Rockefeller Center
1230 Avenue of the Americas
New York, NY 10020

First Touchstone Edition 1998

Designed by Deirdre C. Amthor

Manufactured in the United States of America

3 5 7 9 10 8 6 4 2

The Library of Congress has cataloged
the Simon & Schuster edition as follows:
Rhodes, Richard.
Deadly feasts : tracking the secrets of a terrifying new plague /
Richard Rhodes.
p. cm.
Includes index.
1. Prion diseases—popular works. I. Title.
RA644.P93R46 1997
616.8—dc21 97-320
CIP

ISBN 0-684-82360-8
0-684-84425-7 (Pbk)

Photo credits: Pages 34, 35, 38, 48, 50, 58, 158, 210: D. Carleton Gajdusek. Page 53: Grant Dahmer. Page 93: C. J. Gibbs, Jr.

For Ginger

The author gratefully acknowledges a grant for travel from the Alfred P. Sloan Foundation

CONTENTS

The history of the world, my sweet, is who gets eaten and who gets to eat.

Sweeney Todd

TO THE READER

THE THREAT OF Ebola virus has haunted our nightmares since Richard Preston published his "terrifying true story" *The Hot Zone* in 1994. Ebola hides in the African rain forest, but a deadlier disease than Ebola has begun killing young people in Britain and France. Ebola is a terrorist: it sickens people quickly and spares at least one out of ten. The new disease is a stealth agent: it incubates silently for years and kills every last victim it infects. Ebola is a sickness of fever and bleeding, no worse than cholera, a quick if not a merciful death. The new disease is an atrocity of destruction—a headache, a stumble, and then hallucination, palsy, seizure and coma drawn out horribly for months. Victims' brains go spongy; their minds dim; they lose the ability to walk, to talk, to see, to swallow; they die slowly, drowning in pneumonia, or they starve to death.

Ebola can survive outside the body for a few days at best. Sunlight kills it. Ultraviolet light kills it. The new disease agent refuses to die. Assault with pressurized, superheated steam in the autoclaves that hospitals use to sterilize instruments for surgery barely slows it. It remains deadly after hours of intense bombardment with hard radiation, months of soaking in formaldehyde, years of burial, decades of freezing. It survives even the fiery furnace of a seven-hundred-degree oven.

How Ebola spreads is still uncertain, but scientists know it's a virus. In time, a vaccine will protect us from its threat. The new disease turns up no virus in victims' brains. It creeps past the barriers of species and immunity. Evidence accumulates that it's a bad seed, a mistake of protein, a misshapen crystal that forces the brain to poison itself. If so, it's a new kind of disease agent that can never be eradicated.

How the new disease spreads is known: it spreads in the cannibalism of animals by animals, it spreads in the industrial cannibalism of animal remains fed to animals, it spreads by the eating of beef.

Nothing that you are about to read is fiction. No names have been changed. However harrowing, every word is true.

Richard Rhodes

Part One

Among the Cannibals

First Connection . . .

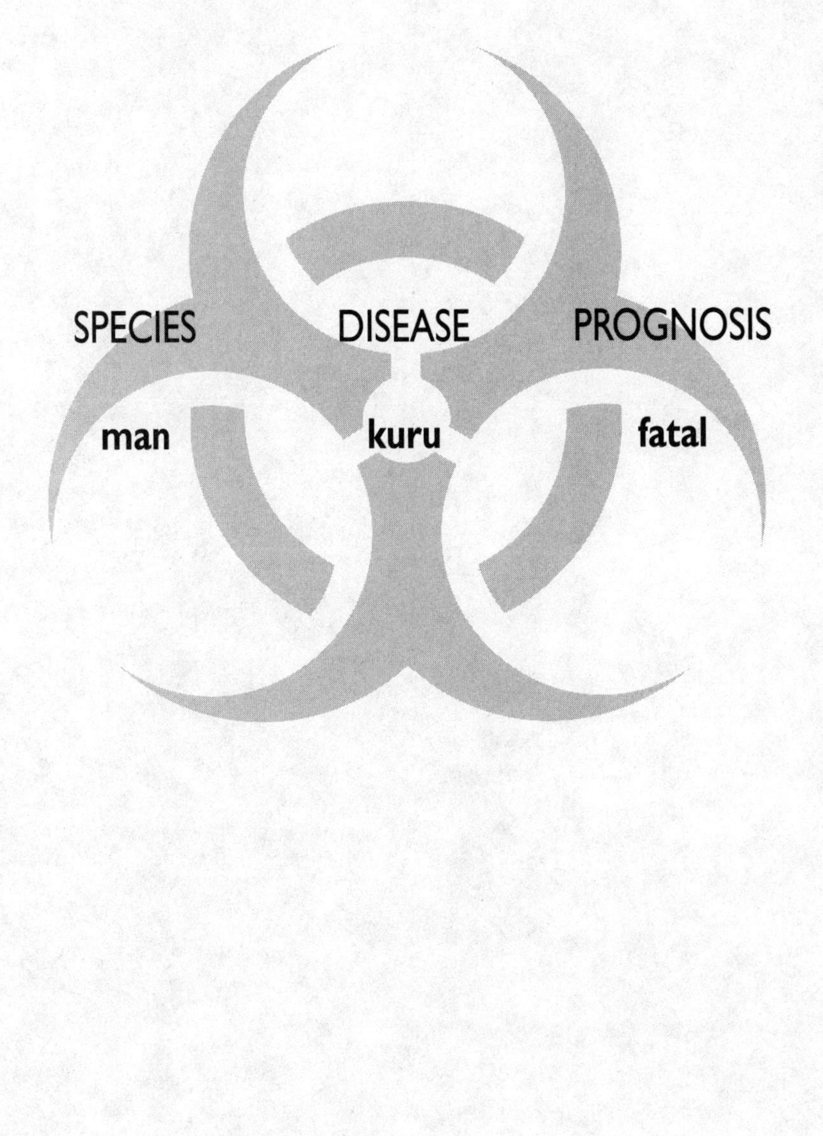

ONE
I Eat You

South Fore, New Guinea Eastern Highlands, 1950

DARK NIGHT in the mountains and no drums beating. No flute music like birdsong from the forest above the village—the men controlled the flutes and this was women's business, secret and delicious, sweet revenge. In pity and mourning but also in eagerness the dead woman's female relatives carried her cold, naked body down to her sweet-potato garden bordered with flowers. They would not abandon her to rot in the ground. Sixty or more women with their babies and small children gathered around, gathered wood, lit cooking fires that caught the light in their eyes and shone on their greased dark skins. The dead woman's daughter and the wife of her adopted son took up knives of split bamboo, their silicate skin sharp as glass. They began to cut the body for the feast.

New Guinea was the last wild place on earth. Its fierce reputation repelled explorers. Micronesians canoe-wrecked anywhere in its vicinity swam the other way. Captain Bligh, put off the *Bounty* after his crew's notorious mutiny, gave the violent island wide berth. It was the largest island in the world after Greenland, fifteen hundred miles long, four hundred miles wide, shaped like a dinosaur with a central cordillera of mountains for a spine, rising out of the Western Pacific just below the Equator north of Australia, eastward of Sumatra and Borneo. Mangrove swamps fouled its tropical coasts; its mountainous interior rose barricaded behind impenetrable, leech-infested rain forests. Its people were Melanesian—small, muscular, black, woolly-haired Stone Age fishermen, hunters and farmers—divided into a thousand warring groups so isolated from each other by conflict and difficult terrain that they spoke more than seven hundred separate languages, one island's cacophony accounting for half the languages on earth.

By the time Dutch, German and English ships began to anchor at the mouths of the island's great tidal rivers, in the mid-nineteenth century, it was common knowledge among Europeans that the savages of New Guinea were cannibals. But there are cannibals and cannibals: warriors who eat their enemies, hating them, but also relatives who eat their kin in a mortuary feast of love. Fore* women ate their kin. "Their bellies are their cemeteries," one observer remarks. "I eat you" was a Fore greeting.

Down in the garden in the flaring firelight, the dead woman's daughters ringed her wrists and ankles, sawed through the tough cartilage, disjointed the bones and passed the wrinkled dark hands and splayed feet to her brother's wife and the wife of her sister's son. Slitting the skin of the arms and legs, the daughters stripped out muscle, distributing it in dripping chunks to kin and friends among the eager crowd of women. They opened the woman's chest and slack belly and the smell of death wafted among the sweet-potato vines. Out came the heavy purple liver, the small green sac of the gallbladder cut carefully away from the underside and its bitterness discarded. Out came the dark red heart gory with clotting blood. Out came the looping coils of intestines, dully shining. Even the feces would be eaten, mixed with edible ferns and cooked in banana leaves.

---

*Pronounced FOR-ae.

They knew the work from slaughtering pigs. They parted out the dead woman by the same kin rights they followed parting out pig, but men got the best parts of pig. Men claimed the small game they shot and trapped in the forest as well, the possums and lizards and the rare flightless cassowary whose legs supplied fine flutes of bone they could wear in their noses. Women grew gardens of beans, sweet potatoes and sugarcane, nourishing but bland, and supplemented their mostly vegetarian diet with roasted hand-sized spiders and fat grubs. Men lived separately from women and children, following their wives into their gardens to copulate, sharing the big men's lodge with the older boys. Men believed contact with women weakened them. They resented the fecundity of women. Men seldom ate the dead and then only the red meat, surreptitiously.

The crowd of women and children got busy at collecting and chopping as the body of the dead woman diminished. (Her name survives as a discreet abbreviation in a medical thesis: *Tom.* Tomasa?) One of the daughters doing the butchering cut around the neck, severed the larynx and esophagus, sawed through the cartilage connecting the vertebrae, disjointed the spine and lifted the head aside. The other daughter skinned back the scalp skillfully, took up a stone ax, cracked the skull and scooped the soft pink mass of brain into a bamboo cooking tube. Their cousins, the North Fore, cooked bodies whole with vegetables in steam pits lined with hot stones, but the South Fore preferred mincing the flesh of the dead and steaming it with salt, ginger and leafy vegetables in bamboo tubes laid onto cooking fires. They ate every part of the body, even the bones, which they charred at the open fires to soften them before crumbling them into the tubes. The dead woman's brother's wife received the vulva as her special portion. If the dead had been a man, his penis, a delicacy, would have gone to his wife.

From the coast, the mountainous interior of New Guinea

looked like a single range of peaks, but Australian gold miners following traces of color—gold dust in the streams—up into the mountains in the late 1920s discovered a vast unknown interior between multiple parallel ranges. High, temperate mountain valleys sustained a population of more than two hundred thousand people in villages and hamlets cleared from primeval pine forests, a true Shangri-La. When the first white men crossed into the high valleys, their arrival stunned the isolated highlanders, who did not even know the ocean existed beyond the mountains that closed them in. Old motion pictures show the highlanders laughing, crying, shuddering with excitement at their first encounter with the white apparitions; they imagined them to be their kinsmen returning from the dead. They knew nothing of guns and showed no fear of them, but ran away in terror the first time an Australian took out his false teeth. They were true democrats and pledged fealty to no chiefs, but they distinguished big men from ordinary men and from men who were nothing. They wore beaded and feathered headdresses, nose bones, necklaces of pig tusks and aprons of woven bark or grass and smeared their bodies with fire char and rancid pig fat against the insects and the cold. Men carried stone axes or longbows. Some of them affected phallocarps instead of aprons—braggadocio penis sheaths made of great curving hornbill beaks or ornate sea shells traded up from the unknown coast. Women wore grass skirts and went bare-breasted. They cut off finger joints in mourning, wore mourning necklaces of the dried hands of lost babies, carried a husband's rotting head in a woven bag, a *bilum,* on their backs for months after his loss, suffering the stink.

Eating the dead was not a primordial Fore custom. It had started within the lifetime of the oldest grandmothers among them, at the turn of the century or not long before. They learned it from their neighbors to the north. It spread to a North Fore village and word got around. "This is sweet," an anthropologist reports the Fore women saying when they first tasted human flesh. "What is the matter with us, are we mad? Here is good food and we have neglected to eat it. In future we shall always eat the dead, men, women, and children. Why should we throw away good meat? It is not right!" The meat was sweet and so was the revenge the women took thereby against the men who claimed the best parts of pig—pigs the women had sometimes suckled at their own breasts. They did not eat lepers or those who died of diarrhea, but the flesh of women killed by sorcery they considered clean. Dying Fore asked to be eaten and assigned their body parts to their favorites in advance.

An anthropologist named Walter Arens published a book in 1979 claiming that cannibalism has always been a myth everywhere in the world, a tale told at third hand not to be believed. To the contrary, cannibalism was a fact everywhere in the primitive world, the Old World and the New, and still sometimes emerges. Fossilized human bones with cut-marks bear witness in European caves across seven hundred thousand years. A dissident Chinese journalist, Zheng Yi, discovered officially sanctioned cannibalism of the livers of class enemies in the province of Guangxi during the decade of Cultural Revolution that began in the People's Republic of China in 1966.

The Fore admitted their cannibalism freely to the first Europeans who questioned them, though they gave it up when missionaries and Australian police patrols pressed them to do so in the late 1950s—Sputnik was beeping overhead—and deny it today. Whatever its connection with ritual, cannibalism in New Guinea was also a significant source of protein, two American anthropologists have calculated: "A local New Guinea group of one hundred people (forty-six of whom are adults) which obtains and eats some five to ten adult victims per year would get as much meat from eating people as it does from eating pork."

The women at their mortuary feast butchered and cooked down in the garden, but they ate in private, carrying the steaming bamboo tubes back to their separate women's houses, sharing the feast with their children. A young American doctor who came a few years later to live and work among them thought their eating habits almost as surreptitious as the toilet habits of Westerners. It wasn't that they were ashamed of eating the dead—they were just as surreptitious with pig. Eating meat was orgiastic. The men said that the women were insatiable, wild, like the forest. When the men pulled the wild grass at the edge of the forest they said it was women's pubic hair. Marriage barely tamed them.

Lately, more and more Fore women had been dying of sorcery, which only men practiced, a fatal bewitchment they called kuru.* Kuru meant shivering—with cold or with fear—and by 1950 it was claiming women in every Fore village. The Fore men earned a fearsome reputation across the highlands as sorcerers. Once the shivers of kuru began, the bewitchment progressed inexorably to death. Women bewitched with kuru staggered to walk, walked with a stick and then could no longer walk at all. Before losing the ability to swallow they got fat and the flesh of those who died early of pneumonia was rich meat.

*Pronounced KOO-roo.

TWO
Kuru

South Fore, New Guinea Eastern Highlands, 1957

WHEN DR. D. CARLETON GAJDUSEK* stopped off in New Guinea in March 1957 on his way home from research work in Australia, he expected to spend a few months exploring what he called "child growth and development in primitive cultures." Arriving in Port Moresby, the thirty-four-year-old American pediatrician and virologist paid his respects first of all to the new director of public health for the Australian trust territory, a physician named Roy Skaggs. Skaggs was delighted to see him. One of Skaggs's public-health officers, Dr. Vincent Zigas, who was based in the frontier settlement of Kainantu in the New Guinea Eastern Highlands and whose district included the North and South Fore, had recently appealed for help studying the strange, deadly disease the Fore called kuru. Zigas didn't believe kuru was caused by sorcery. He thought it might be a new form of encephalitis—an infection of the brain. Skaggs had just returned from visiting Zigas at the hospital Zigas was building beside the Kainantu airstrip and had examined two middle-aged Fore women sick with kuru. The discovery of a new human disease is as exciting to medical researchers as the discovery of a new element

---

*Pronounced GUY-dew-sheck.

is to physicists. Skaggs had no budget for exploring new diseases, but Gajdusek had first-class credentials as a research scientist and he was paying his own way. The public-health director encouraged the American to fly up to Kainantu and take a look.

Gajdusek arrived at the frontier outpost on March 13. Zigas, a trim, volatile Estonian with wavy hair who resembled the comic actor Danny Kaye, recalled in a posthumously published memoir encountering his curious colleague for the first time outside his house:

> My preoccupation was interrupted by a peculiar visitor. At first glance he looked like a hippie, though shorn of beard and long hair, who had rebelled and run off to the Stone Age world. He wore much-worn shorts, an unbuttoned brownish-plaid shirt revealing a dirty T-shirt, and tattered sneakers. He was tall and lean, and one of those people whose age was difficult to guess, looking boyish with a soot-black crew-cut unevenly trimmed, as if done by himself. He was just plain shabby. He was a well-built man with a remarkably shaped head, curiously piercing eyes, and ears that stood out from his head. . . . Everything he possessed spoke of his being peripatetic. Even standing still, he seemed to be on the move, with top tilted forward, in the breathless posture of someone who never had time enough to get where he had to be. . . . I guessed him to be from America, a nation of strange mixtures of blood. . . . I was machine-gunned by his numerous questions.

"At Kainantu there were two old women who had kuru," Gajdusek recorded of that first encounter with Zigas's patients. When he saw the Fore women they could no longer walk. They shivered uncontrollably, not from cold but from damage to their brains. Their arms and legs pulsed as well with slow, continuous, involuntary tremors—"athetoid movements," Gajdusek called them. Their speech was blurred. "They were rational, but articulation of speech was very poor. Silly smiles, with grimacing, were prominent." Zigas would remember the phrase Gajdusek coined for the exaggerated emotionalism of kuru: "pathological laughter." When the Australian press heard of kuru, it would inflate that characteristic to a name for the disease itself and call it "laughing death." Close up, it was anything but funny.

By evening, says Gajdusek, Zigas "had exhausted me in rounds examining his patients, including the two women with kuru. Later that night, he told me more about this fascinating disease among the endocannibal Stone Age Fore people in the 'uncontrolled region' to the south. His enthusiasm for the challenge presented by this new disease, and his recognition that its high epidemic incidence demanded intensive research to determine its [origins], caused me to alter my plans and to arrange an expedition with him to the [South Fore] to see kuru." They left together by jeep the next morning.

Carleton Gajdusek was a Czech-American prodigy, born in 1923 in Yonkers, New York, where his immigrant father owned a butcher shop. As a boy, he had thrilled to the stories of the great biologists and physicians in Paul de Kruif's book *Microbe Hunters* who had revolutionized medicine in the nineteenth and early twentieth centuries. Young Carleton had inscribed the names of the men de Kruif celebrated onto the stairs leading up to the childhood laboratory he assembled in the attic of his family's large house: *Leeuwenhoek,* who first saw microbes in a drop of rainwater through a homemade lens; *Koch,* who first demonstrated that germs are alive and cause disease; *Pasteur,* who invented pasteurization and devised the first laboratory-produced vaccines; *Walter Reed,* who conquered yellow fever. These and a half dozen

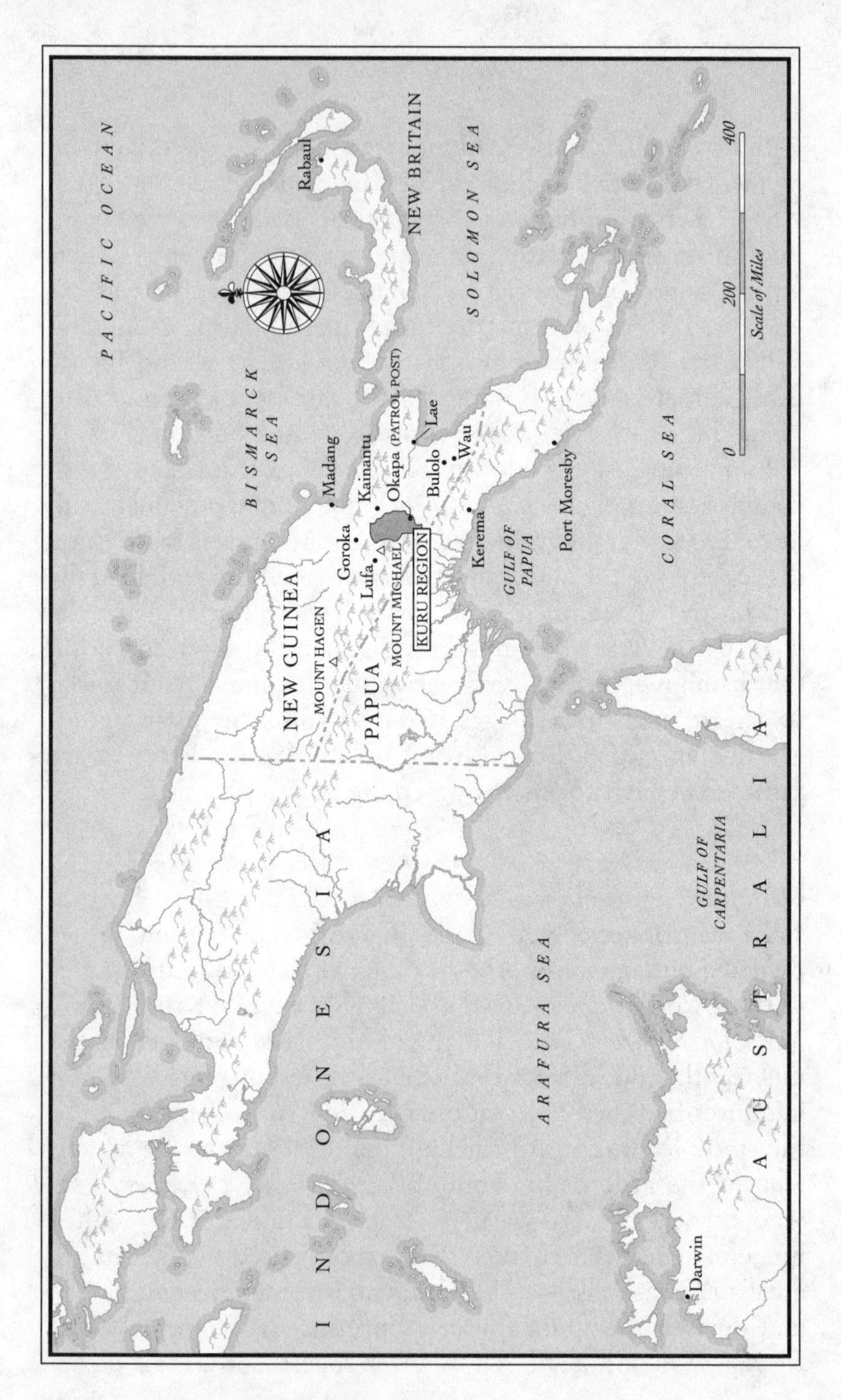

The kuru region of Papua New Guinea.

others from de Kruif's book had inspired the boy to dream of going into science. His aunt was a scientist—an entomologist, a specialist in insects—and he took inspiration from her as well. "She had me doing experiments just after I was a toddler," he recalls—studying the effects of pesticides on insect larvae.

By the time Carleton Gajdusek was a young teen he had begun original research at the institute where his Tante Irene worked. By nineteen he had taken a biophysics degree summa cum laude at the University of Rochester under the distinguished physicist Victor Weisskopf and was entering Harvard Medical School. He chose to specialize in pediatrics, but basic research drew him as well and he went off to Caltech after medical school for a year to study physical chemistry under Linus Pauling, the Nobel laureate chemist. From Caltech, Gajdusek returned to Harvard for postdoctoral work in microbiology under John Enders, whose success in growing viruses in tissue culture—outside the body on beds of nutrient in glass dishes—earned him a Nobel Prize and provided a technical basis for the development of polio vaccine. Nobel laureates pick their assistants from among the most promising young scientists they encounter, passing on their unwritten knowledge of scientific craft; that a succession of Nobel laureates chose to mentor Carleton Gajdusek calibrates the high promise of his gifts.

In those days the young American research physician was rail-thin, intense, enormously ambitious but scattered across a thousand projects; Enders said of him that he "was very bright but you never knew when he would leave off work for a week to study Hegel or a month to go off to work with Hopi Indians." Gajdusek was and still is a compulsive talker who spills ideas nonstop for hours—good talk, often brilliant talk and consummate storytelling, but more than some listeners can bear. A neurologist friend of his speculates that his compulsive flow of speech is a kind of epilepsy, and as a young man he learned that relaxing to a level the rest of humanity considers normal precipitated crippling migraine headaches that laid him low for days. Women found his intensity attractive—and his Slavic features, sensual mouth and mischievous, intelligent eyes—but he submerged himself compulsively in work. Macfarlane Burnet, the Australian Nobel laureate who invited Gajdusek to work with him in Melbourne in the mid-1950s—the work that led the American to New Guinea and the Fore—characterized him shrewdly around that time in a letter to a critic:

> My own summing up was that he had an intelligence quotient in the 180s and the emotional immaturity of a fifteen-year-old. He is quite manically energetic when his enthusiasm is roused and can inspire enthusiasm in his technical assistants. He is completely self-centered, thick-skinned and inconsiderate, but equally won't let danger, physical difficulty or other people's feelings interfere in the least with what he wants to do. He apparently has no interest in women but an almost obsessional interest in children, none whatever in clothes and cleanliness and he can live cheerfully in a slum or a grass hut.

By the time Gajdusek got to Australia, in 1955, he had lived in any number of slums and grass huts—investigating rabies and plague in the Middle East, viruses among North and South American Indians, epidemic hemorrhagic fever in Korea during the Korean War as a captain in the U.S. Army Medical Corps, encephalitis in the Soviet Union. Enders at Harvard had passed him to a tough, smart scientist-administrator at the Walter Reed Army Institute of Research in Washington named Joe Smadel who knew how to put the bit to such thoroughbreds ("Smadel at Washington," Burnet advises Gajdusek's critic, "said the only way to handle him was to kick him in the tail, hard. Somebody else told me he was fine but there just wasn't anything human about him.") Joe Smadel kicked tail. Gajdusek delivered.

Gajdusek spent a year with Mac Burnet in Melbourne. Among other things, he worked on infectious hepatitis and introduced the Australians to viral tissue culture. By June 1956 he was planning his visit to New Guinea, but he stumbled upon a way to develop a new medical diagnostic test related to skin grafting—"the first really original and important thing I have turned up," he reported to Smadel—and that research held him a further six months in Australia. In March he descended on Port Moresby and discovered the work that would occupy him for the next forty years.

Okapa, where Gajdusek and Zigas arrived after four hours of rough driving down from Kainantu with the two crippled Fore women hanging on in the back seat, was a police patrol post set on an open hillside plateau surrounded by pine forest at five thousand feet in the heart of Fore country. It was pouring rain when they arrived and it continued raining for days. Gajdusek immediately collected an entourage of Fore children, who followed him around calling him "Docta America" and "Karton." He kept a rat-a-tat-tat daily journal on a battered portable typewriter, talking to the journal when he couldn't bend a living ear, and the alien machine attracted them like moths to flame. They swarmed over him, passing along menageries of fleas and lice—bathing was not a Fore custom. "I am in one of the most remote, recently opened regions of New Guinea," he wrote Joe Smadel with a swagger two days after he arrived, ". . . in the center of tribal groups of cannibals . . . still spearing each other as of a few days ago, and cooking and feeding the children the body of a kuru case . . . only a few weeks ago."

Kuru was a disease almost exclusively of women and young children. Gajdusek and Zigas set to work examining them:

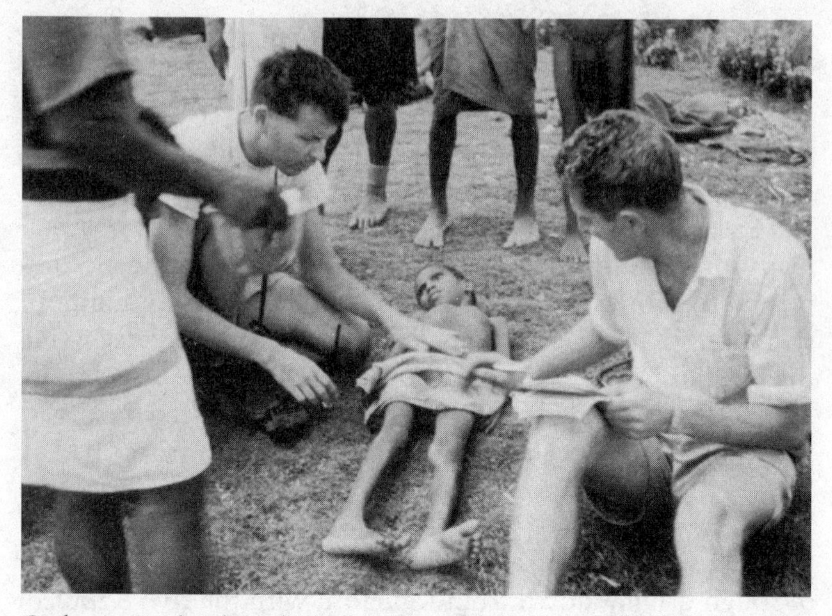

Carleton Gajdusek and Vincent Zigas examining a child kuru victim, Okapa, 1957.

> One child, a boy of about seven, had been carried here; he is obviously unlikely to survive for long, can no longer walk, has hardly distinguishable speech, and urinates and defecates in the house, although he is not dribbling nor incontinent. As with all cases, he has to be fed now, having lost the ability to bring food to his mouth. It is hard to believe that he is a recent case—a boy previously of normal intelligence and physical development—but multiple reliable informants testify that he was walking, running and playing normally only three months ago. . . . The course is that described for classical kuru, i.e., one month of unsteady gait followed by tremors and athetosis and blurred speech in the second month and now in the third month almost complete incapacitation.

Vincent Zigas, patrol officer Jack Baker and Carleton Gajdusek examining kuru brain (in pan on dining table) at Okapa, 1957.

Pidgin had evolved in New Guinea as a common language, a colorful amalgam of English, German and Spanish with a few native words thrown in. The two doctors communicated with the Fore through Zigas's young *dokta boi* assistants, who spoke Fore and Pidgin. Zigas was fluent in Pidgin, and Gajdusek, a natural linguist already comfortable in five or six European languages, quickly picked up the simplified tongue and began learning Fore from his gang of dazzled kids. The men soon realized, as Gajdusek wrote, that "native diagnosis of kuru is as reliable as any modern medical appraisal would be." The Fore divided the progressive and fatal neurological disorder into three stages: *wokabaut yet*—"walk-about yet," still ambulatory; *sindaun pinis*—"sit down finish," no longer able to walk; and *slip pinis*—"sleep finish," stupor. The Fore also accurately identified a preliminary phase, *kuru laik i-kamap nau*—"kuru like he come up now," kuru is about to begin; and a terminal phase, *klostu dai nau*—"close to die now," near death.

"It is so astonishing an illness," Gajdusek wrote Joe Smadel from Okapa, "that clinical description can only be read with skepticism. . . . Classical advancing 'Parkinsonism' involving every age, overwhelming in females although many boys and a few men also have had it, is a mighty strange syndrome. To see whole groups of well-nourished, healthy young adults dancing about, with athetoid tremors which look far more hysterical than organic, is a real sight. And to see them, however, regularly progress to neurological degeneration in three to six months, usually three, and to death is another matter and cannot be shrugged off." Death came cruelly: prostrate kuru victims lost the ability to swallow and starved and thirsted to death, or succumbed before starvation to pneumonia or to the deep, gangrenous bedsores they developed lying in their own urine and excrement, unable to move. Nor did dementia ease their awareness of their suffering; with kuru they usually remained conscious and alert, but speechless, until near the end.

The signs of kuru were like the signs of degenerative brain diseases known in the West—Parkinson's disease, Alzheimer's, multiple sclerosis, ALS. But all those diseases were believed to be degenerative—caused by pathological alterations in the tissues of the brain—not infectious, and none spread as epidemics. In the weeks after the two doctors' arrival at Okapa, as they conducted examinations and went out patrolling the South Fore with a line of carrier *bois* on survey—cutting trails through rain forests resplendent with orchids and bright orange impatiens, wading the deep red mud—they discovered that every Fore village and hamlet had a history of recent kuru deaths. Their examinations soon extended to a series of some two hundred patients—fifty children, almost all the rest adult women. The disease, they calculated, annually afflicted fully one percent of the total Fore population of about thirty thousand, some three hundred cases a year. In the worst-hit Fore settlements the incidence reached five to ten percent, and more than half of all deaths in the past five years in those settlements had been from kuru. It was, Gajdusek wrote Smadel grimly, "the major disease problem among the Fore people, and beyond infancy . . . it is perhaps the major cause of death next to warfare wounds."

Zigas had assumed that the disease was caused by an infection, and so at first did Gajdusek. But infection—the alien protein of foreign organisms invading the human body—causes inflammation as the body's lymph and immune systems respond defensively to destroy the invader. With inflammation come fever, increased numbers of lymph cells in the cerebrospinal fluid that bathes the spinal cord and the brain and other physical changes. To their astonishment, the two doctors found none of these signs in their kuru patients. "It's hard to explain to a nonmedical person how strange it is not to see inflammation," Gajdusek told me the first time I interviewed him. "And with kuru there was *none.* Absolutely none."

Yet some disease process, not sorcery, must have been causing the disease, whatever the Fore believed. The two men took cultures from the kuru-afflicted and sent them to Port Moresby to see if the laboratory there could grow the disease organisms, but the cultures turned up nothing out of the ordinary. Could the disease be hereditary, passed down on the female side from mother to child? If so, Gajdusek wrote Smadel, "we have the highest-incidence concentrated 'epidemic' [of hereditary disease] ever seen." Zigas and Gajdusek decided competent neuropathologists needed to section and study the brains of some of their kuru patients, where the damage obviously occurred. The patients would soon be dying. A way would have to be found to collect, preserve and transport whole brains from Okapa to Australia and to the United States. They would have to do autopsies on Stone Age cannibals in the Eastern Highlands bush.

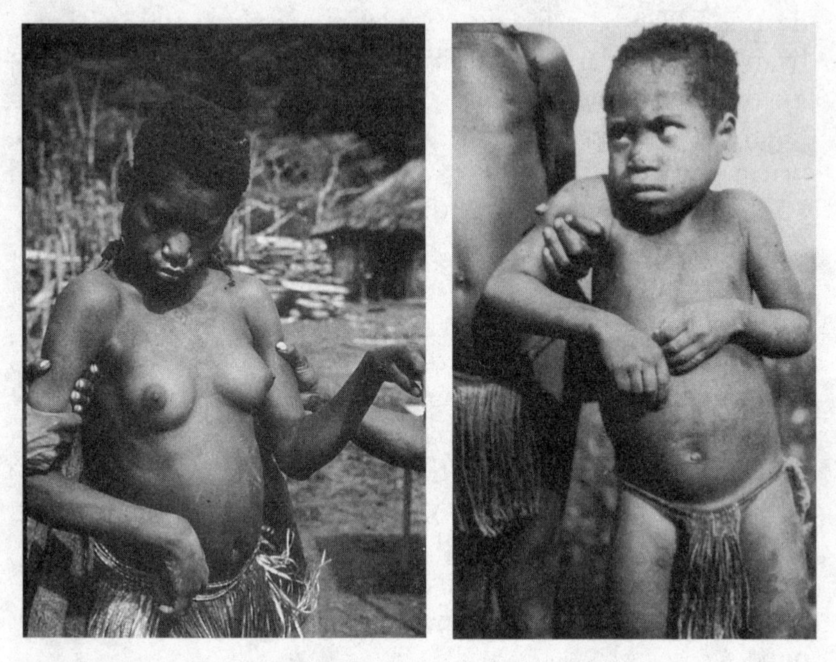

Two kuru victims, a young Fore woman and a small boy.

Their first opportunity came in the middle of May:

> Dear Joe,
> . . . I write at the moment to let you know that we have had a kuru death and a complete autopsy. I did it at 2 a.m., during a howling storm, in a native hut, by lantern light, and sectioned the brain without a brain knife. But the brain in 10% Formol saline is off to Melbourne for neuropathology, along with pieces of all organs I sampled[:] Liver, kidney, muscle and the fascia of anterior abdominal wall, lung, pancreas, aortic wall,

> section of cervical cord, ovary, a piece of cerebral cortex, and the spleen. . . . The next autopsy I shall try to get off to you in entirety, rather than to Melbourne.

The first brain went to Melbourne because the Australian press had picked up the "laughing death" story and Mac Burnet had expressed his displeasure that native-born Australians weren't handling the investigation (Zigas was a refugee from Estonia who had found his way to Australia after the Second World War). Gajdusek brashly wrote his Australian mentor to lay off. "Since then [Burnet] has been most cooperative," he bragged to Joe Smadel. By now Zigas had returned to his Kainantu duties. Zigas made the difficult drive to Okapa whenever he could get away, but Gajdusek carried the work.

He was throwing every treatment he could think of at the disease. "I have not yet seen any response to antibiotics, phenobarbital, Benadryl or Pyribenzamine, cortisone acetate, ACTH, aspirin or vitamin preparations," he wrote Smadel. "We have had a remarkable chance to evaluate these potential therapies, but our patients have degenerated while on high doses. . . . If you can get any of the tranquilizers to me . . . or more particularly, any of the new anti-Parkinsonism drugs, I should be most anxious to try them. . . . I have the native-material house nearly finished and we will be able to move our lumbar puncture, autopsy and urinalysis and reagent preparations off . . . the local patrol officer's dining table shortly." Gajdusek and the patrol officer had organized a great gathering of nearly a thousand Fore men to build "a majestic new hospital" of native materials as well as a smaller house for Gajdusek's use; the thatched house was "quite a cozy place in the stormy downpour which beats on our mountaintop." But the Fore resisted his autopsies. "Our excannibals (and not 'ex') do not like the idea of opening the head, although other dismemberment does not seem to perturb them . . . death, however, away from their remote villages does!" They were "not people you can push. They are proud and have their own ideas, which are most intelligent, and although they have conceded that I can cure their meningitis and pneumonia, they have decided that this [kuru] magic is too strong for me and that my prolonging life by treating and controlling [bedsores] is no blessing at all. They want to die at home, and once fully incapacitated, they want to die as quickly as possible. With such apparently hopeless neurological disease, you cannot blame them."

Then a fifty-year-old woman named Yabaiotu died of kuru at the hospital and her relatives allowed Gajdusek to take her brain. No pathologist, he had compromised the first autopsy by cutting the fresh brain into sections, mashing it badly—fresh brain has the consistency of soft scrambled eggs—but Smadel had sent brain-handling instructions just before this second postmortem came available and Gajdusek followed them carefully: he set the whole brain to pickle for a fortnight in formaldehyde solution, wrapped the now-rubbery organ in gauze and cotton and dispatched it via bush pilot and Qantas Airways around the world to Smadel.

Joe Smadel in the meantime had moved from Walter Reed to an associate directorship at the National Institutes of Health in Bethesda, Maryland, outside Washington, D.C., the United States' premier institution of public medical research. He controlled important resources, and he proceeded to draw on those resources to facilitate Gajdusek's work. He solicited a grant of a thousand dollars from the National Foundation for Infantile Paralysis for his protégé—Gajdusek was running out of personal savings—and he started lobbying medical journals to publish a first report on kuru that Gajdusek and Zigas had drafted and sent along to establish their priority in discovering the disease. Most significantly, he arranged for a first-class neuropathologist at the NIH, Dr. Igor Klatzo, to begin examining kuru brains.

Yabaiotu's brain was the first of a series that Gajdusek shipped from the wilds of highland New Guinea to Klatzo's pathology lab at the NIH in Bethesda. Five more brains followed in the course of the summer, several of them from children. In a preliminary report on the woman Yabaiotu in mid-August, Klatzo found extensive damage to the part of the brain that controls the timing of muscular activity, the cerebellum. Nerve cells in the cerebellum and elsewhere in the lower regions of the brain had degenerated, but Klatzo wrote that his findings "do not give any ground to suspect [an] infectious or inflammatory process." On the other hand, "the widespread involvement of various [brain] structures does not fit into any of the known hereditary degenerative patterns" either. Klatzo thought the condition might be caused by some toxic substance the Fore were encountering in their food or their environment.

Gajdusek had been looking into just that possibility, so far without success. An anthropologist the Australians had assigned him was making exhaustive studies of the Fore diet; she found nothing in the long lists of plants and animals she compiled in her interviews to distinguish it from the diets of their neighbors.

Not all the Fore's neighbors ate their dead, however. The possibility that Fore cannibalism was implicated in the transmission of kuru had been obvious from the beginning. An Australian physician, Michael Alpers, who joined Gajdusek a few years later in New Guinea, told me he remembers hearing the theory spouted in the bars in Goroka, the Eastern Highlands district capital: "Some old prospector would say, 'Oh, those people out there, I mean they get this terrible disease.' And then he'd stage-whisper, *'Well, of course they do, they eat each other, they're all bloody cannibals.'*" But no evidence of inflammation meant there was no evidence of infection, which implied there was no infectious agent—no bacterium, no virus, no Rickettsia or other parasite—for the cannibalism to spread, and indeed none had turned up in the specimens Gajdusek had cultured. And however repugnant to Westerners the idea of eating human flesh, such a grisly diet was not toxic in and of itself, or other New Guinea cannibals would presumably also have sickened with kuru. To the contrary, with rare exceptions—a few women from other groups who married into the tribe—kuru was confined to the Fore. So however baffling the facts, their logic compelled Gajdusek and his colleagues to set the cannibalism theory aside. "I am ever more convinced," he wrote Joe Smadel that summer, "that kuru is one of the most exciting and fascinating medical problems unsolved today, and one which offers to medicine many an insight, especially in this modern age of fear of gene-induced dangers to populations . . . for where in the world is another population so threatened by a lethal set of genes? Obviously, I am still thinking genetically, but remain open to toxic, infectious and other possibilities, if they can be found. No luck yet."

Klatzo continued his postmortem examinations through the late summer and reported again in September. He still thought "some toxic metabolism" was responsible for the damage he identified in the brains Gajdusek sent him. But this time around, studying the brains of dead children, he saw a type of damage that reminded him of an existing human disease. He shared his association with Gajdusek. The damage he found was the presence of microscopic knots of a form of protein known as amyloid. These "amyloid plaques" had never before been seen in a child's brain. They were a familiar by-product of aging and they were common in the brains of people afflicted with Alzheimer's disease. The amyloid plaques Klatzo found in the brains of children dead of kuru were particularly large. In the photographs Klatzo took through his microscope they looked like black, hairy disks floating among the smaller gray nerve cells. Later, when Gajdusek had a chance to talk to Klatzo in person, the two physicians joked that the florid kuru plaques represented "galloping senescence of the juvenile"—that is, accelerated aging. But neither man attached much weight to the unique finding of amyloid in the brains of kuru children, because in some of the cases the condition didn't show up in the sections of brain Klatzo examined. "But of course," Gajdusek says today, rueful of missed clues, "you don't section the entire brain, only sample portions."

Amyloid plaques and the other characteristic damage Klatzo observed in the kuru brains reminded him of a rare degenerative disease that two German doctors had first described nearly four decades before. "The closest condition I can think of," he wrote Gajdusek, "is that described by Jakob and Creutzfeldt." It wasn't quite the same, since it affected the higher regions of the brain more extensively than kuru did and was typically a disease of middle age—"no cases involving children or adolescents are mentioned." Like kuru, it was of unknown cause. It was seemingly much rarer; "only about 20 cases have been reported." In conclusion, Klatzo thought that there were "no known hereditary disorders resembling even remotely the kuru condition." Nevertheless, the damage kuru caused to the brain was similar to the damage caused by the rare condition known as Creutzfeldt-Jakob disease. That was a slim lead, but it was more than Gajdusek had before. In the years ahead, he would widen it to a cascade of new discoveries.

Second Connection . . .

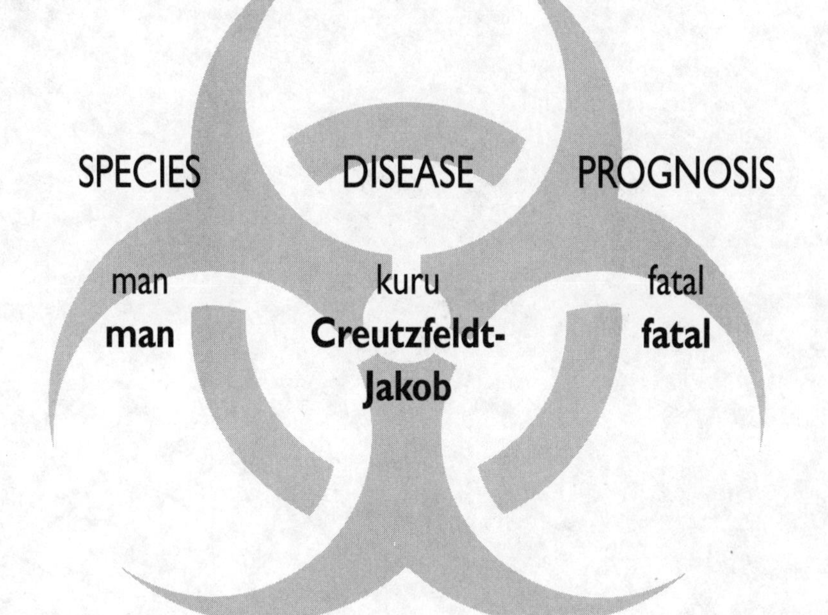

THREE
Dr. Creutzfeldt and Dr. Jakob

Breslau and Hamburg, Germany, 1913–1921 / London, 1959

IN JUNE 1913, a maid in a convent in Breslau, Germany, named Bertha Elschker suffered a breakdown and was taken to a clinic at the University of Breslau directed by the renowned neurologist Alois Alzheimer, the discoverer of Alzheimer's disease. One of Alzheimer's assistants, a physician named Hans Gerhard Creutzfeldt, recorded the twenty-three-year-old orphan's history. Her normally cheerful personality had changed abruptly about a month before. "She no longer wanted to eat or to bathe," Creutzfeldt noted, "she neglected her appearance, became dirty . . . [and] assumed peculiar postures." Three days before he saw her, "she suddenly screamed out that her sister was dead, that she was to blame, that she was possessed of the devil, that she herself was dead, that she wanted to sacrifice herself." These changes might have been signs of mental illness, but Bertha's dazed expression, silly giggling, twitching eyes and unsteady walk made it obvious to Creutzfeldt that she was suffering from physical damage to her brain.

The young German doctor examined the emaciated woman and found fluttering facial muscles, ticlike jerks in her arms, tremors that started up whenever she made a voluntary movement (a sign of neurological damage that doctors call intention tremor), altered reflexes, distracted speech and what Creutzfeldt judged to be "unmotivated outbursts of laughter." He admitted her to the hospital. Her physical condition progressively worsened and her mental condition deteriorated. Some days she screamed continually. Other days she languished in a stupor. By the beginning of August she had sunk into status epilepticus—one epileptic seizure following another in rapid succession—and responded only if Creutzfeldt pricked her with a pin. He chronicles her last days: "On August 6 a genuine epileptic attack occurs. . . ; towards evening, a second attack. . . . In the ensuing days, the patient lies . . . twitching. . . . In the last hours, stupor deep-

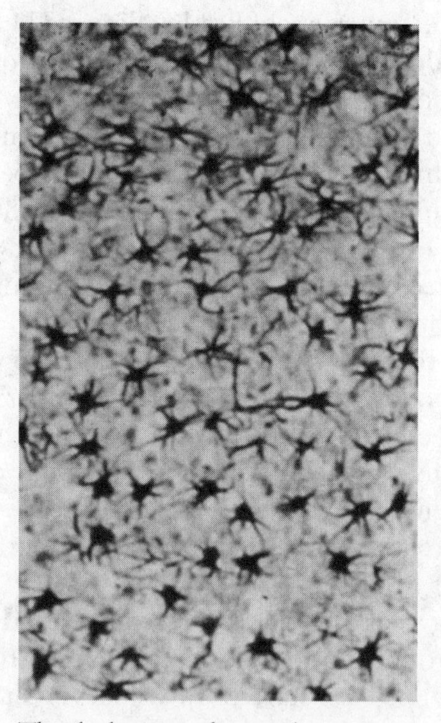

The dark stars of astrogliosis.

ens, swallowing is impaired; death ensues on August 11, in status epilepticus."

After Bertha's death, Creutzfeldt autopsied her and examined her brain. He found extensive brain damage without inflammation. Something had killed millions of brain cells. They had been cleaned away and partly replaced by glial cells, the brain's sometimes destructive repair units (*glia* is Greek for glue). When Creutzfeldt stained cross sections of Bertha's brain to highlight the damage, the swollen, proliferated glia looked under his microscope like brown stars crowding a dead gray sky.

Creutzfeldt recognized a new disease in Bertha Elschker's fatal illness, but the Great War intervened before he could report his discovery. In 1920 he was finally able to prepare a paper for a German medical journal. A colleague at the University of Hamburg, Dr. Alfons Jakob, read the paper while it was still in press. Jakob had lost four patients whose symptoms and autopsy findings seemed to him to match Creutzfeldt's case, although all four were significantly older. Jakob described them in a paper published in 1921. Only one case is still considered canonical, a forty-two-year-old salesman named Ernst K. whose illness began with aching legs and dizziness and progressed like Bertha's to a final stage of dementia and stupor. Ernst K.'s brain also showed no inflammation but extensive cell death and starlike, proliferated glia.

Together, these first early reports described a new degenerative disease of the brain, Creutzfeldt-Jakob disease. CJD is known today to be an uncommon but not a rare disease—rabies is rarer—but for the next four decades very few cases of CJD were reported. When Igor Klatzo associated kuru with CJD in his letter to Carleton Gajdusek in 1957, he knew of only about twenty such reports, none of them American, even though CJD occurs throughout the world with uncanny regularity: similar prevalence and incidence, Gajdusek has written, "in all races and in all climes from the Arctic to the tropics"—about one case per million people worldwide. Few physicians reported CJD in those early days because the disease was easy to confuse with a tangle of other ill-defined brain diseases which no one yet understood.

Creutzfeldt and Jakob, it seems, had left out of their detailed descriptions of brain pathology the truly distinctive mark of the disease that came to bear their name: holes. Under the microscope, the bodies of nerve cells in Bertha Elschker's and Ernst K.'s brains had been as full of holes as a sponge. "Spongiform change," this condition came to be called, and it would prove to be the key to CJD diagnosis. Why did neuropathologists examining CJD brains ignore it down through the decades? Dr. Paul Brown, a world authority on CJD and one of the talented scientists Gajdusek re-

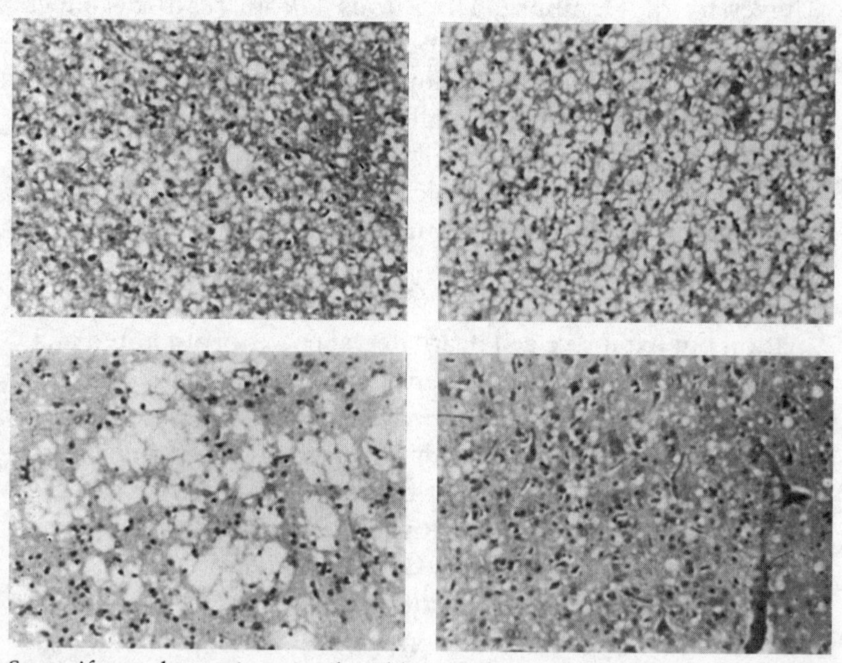

Spongiform change in animal and human brain.

cruited to work with him at the NIH when he returned from New Guinea, believes they did so out of professional embarrassment—"because of an excusable reluctance to attribute significance to holes." Brown, a lean Harvard M.D. with a dry sense of humor, observes that "pathologists abhor a vacuum, they abhor holes and to base a diagnosis on holes is asking a lot of a pathologist. Because they can do things to tissue"—accidentally during the pickling process, for example—"that give them holes that are virtually indistinguishable from CJD. And so for decades these were considered to be unimportant."

When Gajdusek traveled the Fore locating and documenting kuru cases, he was looking for patterns of diet, of incidence, of contact with carriers such as ticks or birds that might help him identify the cause of the disease. This sophisticated detective process is called epidemiology, meaning knowledge of epidemics. "Epidemiology is really the study of distributions," Paul Brown told me. "That's all it is. It's the distribution of something and the clue we look for is unequal distribution. If we're smart or shrewd we can pick out certain groups in the human population that in principle should be at higher risk." The Fore were such a group where kuru was concerned, and within that larger group women and children were at higher risk than men. By contrast, CJD's uniform worldwide distribution has made identifying its origins difficult.

Kuru and CJD caused similar brain damage, but their courses and their epidemiologies were different. "Kuru was virtually a stereotyped illness," says Brown of its symptoms. "CJD runs a gamut from people who go blind and lose their minds to people who have massive incoordination and personality changes as their major features. There's a vast range with CJD. It's really a symphony instead of a melody." Despite the youthfulness of Creutzfeldt's original case, CJD is normally a disease of middle age, rarely showing up before forty years of age and typically occurring in the fifty–to–seventy-five age group, with an average age at death of about sixty years. There's a curious cutoff in frequency after seventy-five. No one knows why. One characteristic both diseases had in common was a puzzling absence of inflammation. To neurologists dealing with CJD, as to Gajdusek with kuru, this lack of an inflammatory response argued against an infectious origin.

Klatzo's connection added to what Gajdusek knew about kuru, but it did not immediately resolve the question of cause. Gajdusek was left to ponder what might link the two obscure, fatal brain disorders and why only one of the two typically killed children.

The human brain was still a medical frontier in the sixth decade of the twentieth century. As long ago as classical Greece, physicians had known that the brain was the center of thought and feeling, but the miniature scale and transparency of its components had defeated anatomists until the perfection of the microscope and the development of tissue staining late in the nineteenth century. Anatomists learned by trial and error that different dyes, each a small discovery of its own, colored different parts of the brain's complex anatomy and made them visible. Applying the dye Congo red to slices of brain, for example, stains amyloid plaques. When a slice stained with Congo red is illuminated under polarized light, any amyloid plaques that might be present stand out edged with a greenish glow.

The brain is the most complex single organ that evolution has yet devised, more than a half million miles of nerve fibers woven together into a soft, deeply folded three-pound mass that floats in nourishing fluid inside a protective container of tough membrane surrounded by the thick bony helmet of the skull. When Carleton Gajdusek was a boy, the brain was usually compared to a central telephone switchboard. After the Second World War, the development of the digital computer suggested a more realistic model. The physicist Philip Morrison once described the brain wryly as "a slow-clockrate modified-digital machine with multiple distinguishable parallel processing, all working in salt water." The brain processes information in the form of electrochemical signals transmitted through nerve fibers that are insulated from each other by fatty glia. The fibers are elaborately interconnected, allowing different areas of the brain to address each other and interact.

Most people are familiar with the cerebrum—the large, walnutlike upper hemispheres of the brain that organize

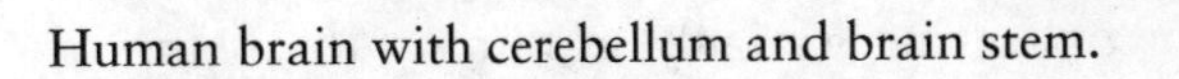

Human brain with cerebellum and brain stem.

thinking, sensory processing and memory. Less familiar is the smaller but life-sustaining cerebellum, the "little brain" that attaches to the brain stem below and at the back of the cerebrum like a hair bun, an organ about the size of a jumbo egg which one researcher calls "[the] most spectacular structure of the vertebrate brain." The cerebellum is the part of the brain that kuru and Creutzfeldt-Jakob disease primarily damage. A single slice through a pickled cerebellum, properly stained to reveal the multiple holes of spongiform change, can definitively establish the presence of either disease. The cerebellum operates unconsciously. Unlike kuru, CJD also usually damages the conscious operations of the cerebrum as well, which explains why CJD victims typically show early signs of dementia (Bertha Elschker's devil-possession, for example) while kuru victims remain alert and aware until near the end.

The cerebellum is the only part of the brain that isn't divided into halves. That distinction hints at its function. So does its astonishingly continuous and uniform structure of precisely spaced parallel nerve fibers and the fact that the entire body is mapped three times from three mutually perpendicular angles onto its wrinkled, banded surface. So far as anatomists can tell (its functions are still debated), the cerebellum is responsible for balance, for muscle tone and for timing the coordination of voluntary movements. Signals that come down from the cerebrum or up from the body pass through the cerebellum, which slows them down or speeds them up so that they work in synchrony. Balance, for example, requires sensing the body's position and unconsciously adjusting arms, legs and torso to counterbalance against tipping backward, forward or side to side—and balance is one of the first senses that kuru and CJD affect. To coordinate complex movement such as writing, speaking, playing a musical instrument or operating a machine, timing the firing of many different large and small muscles is crucial. The cerebellum can resolve intervals of time down to about one ten-thousandth of a second, which is the range of tuning that fine muscle control requires. Kuru and CJD probably affect speech not by damaging the higher speech centers of the cerebrum directly but by throwing off the complex timing necessary to coordinate breath, larynx, palate, tongue and lip control. Creutzfeldt commented in his report on Bertha Elschker that her "unmotivated outbursts of laughter" gave him "the impression of being purely a motor activity." The "pathological laughter" Gajdusek found in kuru probably originated in the same kind of damage. Staggering, falling down, loss of muscle control including the inability to swallow that led to death: all resulted from damage to the cerebellum. But what caused that damage in kuru and CJD was still anyone's guess as the decade of the 1950s approached its end.

By then, Carleton Gajdusek and Vincent Zigas had staked their claim to priority in the discovery of kuru in a number of different medical and scientific journals and Gajdusek had been appointed a visiting scientist at the NIH's National Institute of Neurological Disorders and Stroke. Joe Smadel had recommended the appointment, writing that Gajdusek was "one of the unique individuals in medicine who combines the intelligence of a near genius with the adventurous spirit of a privateer." In the spirit of adventurous privateering, Gajdusek had hiked fifteen hundred miles up and down steep, rain-forested mountains in the Eastern Highlands of New Guinea to define kuru's range, had produced, directed and narrated a medical film about kuru with the help of an Australian film crew, had taken hundreds of blood samples and medical histories and thousands of still photographs of kuru patients and had sent more brains back to Bethesda.

All this strenuous activity had not gone unnoticed. The Australian press had lionized him and *Time* had carried a reasonably accurate report on his work. Gajdusek returned to the United States during 1958 and set up shop at the NIH. He organized an exhibit on kuru, including a large display of color photographs of brain pathology, which began traveling to medical museums in the United States and abroad. In May 1959, he led a discussion on kuru at an international conference in Antwerp, Belgium. By June he was back in New Guinea again, exclaiming in his journal, "At last . . . as though I were returning home!"

At Kainantu in late July, Gajdusek received a remarkable letter from England that changed the direction of his research. The author of the letter was an American research veterinarian named William J. Hadlow, whose specialty was veterinary pathology. For the past year, Hadlow had been working at the British Agricultural Research Council field station at Compton, southeast of London in Berkshire, studying an old but still mysterious disease of sheep.

Born and raised in Ohio, Bill Hadlow was thirty-eight in 1959, two years older than Gajdusek, a trim, handsome man with a high forehead and a square chin, smart and meticulous and as reserved as Gajdusek was flamboyant. Hadlow was on loan to the British from an outpost NIH laboratory in Hamilton, Montana, in a beautiful valley paralleling the magnificent Bitterroot Range of the Montana Rockies. The Rocky Mountain Laboratory had been established by the State of Montana in the 1920s to study the deadly tick-borne disease known as Rocky Mountain spotted fever. There were ticks all over Montana, but the people of Hamilton had been so worried that the ticks the lab was studying would escape and attack the town that its builders had decided to surround the red brick two-story building with a little moat—an early and largely cosmetic version of a biological containment system. By the time Hadlow had arrived, in 1952, the moat had been filled in and the laboratory had passed to the NIH.

Hadlow was uniquely qualified to explore both animal and human pathology. After earning his doctor of veterinary medicine degree at Ohio State in 1948 and working a summer for a vet in upstate New York artificially inseminating dairy cows, he had accepted an invitation from the University of Minnesota to start a department of pathology at the university's new veterinary college. "That was expecting a lot of a recent graduate," Hadlow told me when I interviewed him in Hamilton in 1995. Animal pathology was not yet a veterinary specialty in the U.S. in 1948; what Hadlow really wanted was graduate training in the field himself. Soon after he arrived at Minnesota, he marched across campus to the medical school, found its professor of pathology at his microscope smoking a cigar and convinced the man to allow him to study human pathology along with the medical students.

Hadlow proceeded to take every course the department offered. He got practical experience as well. "Had to help with surgicals that were sent in," he remembers—meaning studies of tissue removed in human surgery. "Had to help on autopsies and in those days everything was funneled into the department. There were just a few pathologists in Minneapolis and St. Paul and they did many more autopsies than they do now. They were doing maybe sixty or seventy percent of the autopsies of persons who died all over the Twin Cities. They sent you out with a medical resident to some cold mortuary in the basement of a little hospital to do a postmortem. I can remember riding a streetcar home with a Mason jar with pickled tissue in it. I didn't even have enough sense to put it in a paper bag. But I got that training. I never had any formal graduate training in veterinary pathology. It was in humans. I had to learn to do animals on my own."

The disease of sheep that Bill Hadlow went to England in 1958 to study was called scrapie. The name describes the most obvious symptom: sheep infected with the disease itch

Sheep afflicted with scrapie.

so intensely that they scrape the wool off their flanks rubbing themselves against walls, trees and fences for relief. Infected sheep also stagger, develop tremors, go blind, fall down and eventually die. What caused the disease and where it originated was obscure. It seemed to be unique. The earliest English record of its appearance dated from 1730, when infected sheep were reported in East Anglia, but it was already prevalent by then in Central Europe. It may have arrived there with the fine Spanish merino sheep that were traded northward from the Middle Ages onward to improve the quality of European and British wool. Epidemic outbreaks of scrapie in the nineteenth century led European authorities to quarantine and slaughter infected flocks, but British sheepmen, many of whom believed the disease was hereditary, hid their losses and the British government neglected scrapie control. By the beginning of the twentieth century, scrapie was endemic in Britain, affecting about one percent of adult sheep annually. Since people had been eating lamb and mutton from scrapie-infected herds for hundreds of years, it was evidently not a disease that spread to humans, but it caused serious losses to farmers.

In the 1930s, French veterinarians demonstrated that scrapie was infectious. They took tissue from a sheep afflicted with natural scrapie, homogenized it and injected it into healthy animals. After a long incubation period, about thirty percent of the healthy sheep began showing symptoms of the disease and eventually died. The French researchers repeated the experiment with goats, which proved to be one hundred percent susceptible: every injected goat came down with scrapie and died.

At about the same time as this French research, the director of the field station at Compton where Hadlow would later come to work, Dr. William Gordon, accidentally conducted a large-scale scrapie transmission experiment in Britain. Gordon and his staff developed a vaccine for a tick-borne sheep virus called louping ill (the virus induced brain damage that caused sheep to spring up and down when they walked, an odd gait the British call louping). The vaccine consisted of homogenized brain, spinal-cord and spleen tissue taken from sheep infected with louping ill, diluted in saline solution and inactivated by adding a small amount of formaldehyde. Compton produced forty-four thousand doses of louping-ill vaccine in 1935 and 1936. In 1937, to Gordon's horror, the sheep inoculated with the 1935 batch of louping-ill vaccine began developing scrapie. The vaccine was evidently contaminated. "I visited most of the farms on which sheep had been vaccinated in 1935," Gordon would recall. ". . . I shall not forget the profound effect on my emotions when I . . . was warmly welcomed because of the great benefits resulting from the application of louping-ill vaccine, whereas the chief purpose of my visit was to determine if scrapie was appearing in the inoculated sheep." Only scrapie's long incubation period saved Gordon's reputation among sheep farmers. Since many of the animals had been three- and four-year-olds at the time of their louping-ill inoculation, most of them had been culled and slaughtered before scrapie could reveal itself. One lesson the vaccine accident taught was that scrapie is tough. Formaldehyde is a powerful antiseptic commonly used for embalming. It killed the louping-ill virus in the vaccine preparation, but the scrapie agent that contaminated the material lived on.

Scrapie appeared in the United States for the first time in 1947, on a sheep farm in Michigan where purebred Suffolk sheep that had originally come from Britain had been imported from Canada. There were larger outbreaks in California in 1952 and Ohio in 1954. The U.S. Department of Agriculture, Hadlow told me, "moved in with a slaughter program": when scrapie appeared in a herd, USDA veterinarians ordered the entire herd slaughtered as a way of containing the disease. The agency also embargoed British sheep imports. "That sparked a lot of interest on the part of the British to do some more research." There were no laboratories in the U.S. set up to study scrapie where American researchers could work. British and U.S. research interests obviously coincided, and the British had the laboratories. The USDA found money left over from the Second World War—a program funded to insure that food procured for U.S. troops abroad was safe—to support the work. "But they didn't have anyone who had any experience with scrapie," Hadlow notes. The agency decided to send someone to Compton for two years to study the disease. That was Bill Hadlow's ticket.

Hadlow remembers Gordon running Compton "as his own little private estate." It was vast by British standards, more than two thousand acres, with three dairy farms and pig and chicken farms as well. "Ostensibly these were to provide experimental animals, but old Bill Gordon was a parsimonious Scotsman and he was running a regular enterprise there"—selling milk and meat on the side. "He was a little disappointed when I arrived because he was expecting a virologist. I said, well, I'm not a virologist, I'm a pathologist and that's what the USDA said they needed here." Hadlow and his wife took up residence in the manor house.

"I thought I could best contribute to the British effort by looking at brains," Hadlow remembers, "which otherwise were usually discarded." The diagnostic hallmark of scrapie was cerebellar holes. To confirm a diagnosis, British pathologists would look at one particular section of a sheep or goat cerebellum, identify spongiform degeneration and throw the rest of the brain away. Hadlow decided to see what other damage scrapie caused to animal brains. Gordon gave him a skilled assistant. "He was first-class. I was fortunate in having him. We used to cut whole cross sections of sheep and goat brains. He had this great big sledge and would whack off these big slices. Not the usual cheeseparing bits that would take half a day just to find out where you were. I got beautiful sections."

Studying his beautiful sections under the microscope, Hadlow identified other characteristic scrapie brain damage besides cerebellar holes. Nerve cells were shrunken or had fallen away. The characteristic sponginess permeated not only the cerebellum but also the outer shell of the cerebrum—the cerebral cortex, the part human beings think with. Most impressive, and previously overlooked in scrapie studies, was the proliferation and abnormal multiplication of star-like glial cells—astrogliosis, the phenomenon is called. "After a year of watching the disease evolve clinically in many animals," Hadlow summarizes what he learned, "of looking at many of their brains, and of absorbing scrapie lore from sundry sources, I thought I had a good idea of what scrapie is like: a protracted degenerative disease of the brain, not an inflammatory one, caused by an infectious agent best thought of then as a virus. I did not know of another disease like it in man or animal."

A friend of Hadlow's, a colleague from the Rocky Mountain Laboratory named Bill Jellison, stopped by at Compton in late June 1959 on his way home from a scientific meeting in Eastern Europe. "At dinner that evening he casually mentioned I might be interested in an exhibit he saw the previous day at the Wellcome Medical Museum in London. It had to do with a strange brain disease of a primitive people in New Guinea." The exhibit was Carleton Gajdusek's kuru show, still making the rounds. Hadlow was curious. At the beginning of July he took a train to London to see the display for himself.

He found it on the first floor of the museum, just inside the main entrance. The large color photographs of the Fore—the people and the place—were interesting, but Hadlow was riveted by the color photomicrographs of brain sections full of holes. In his experience, holes were unusual in human brains, and they reminded him immediately of the spongiform degeneration he had been studying for the past year in sheep and goats. He copied down the titles of several of Gajdusek and Zigas's papers. Back at Compton he wrote Gajdusek at the NIH asking for reprints. Gajdusek's secretary sent them along early in July, and Hadlow read them with fascination, struck by the similarities between the two diseases. "I found the overall resemblance of kuru and scrapie to be uncanny," he recalls. He decided that the similarities were potentially too important to keep to himself. It took him a week to draft a careful letter to the editor of the prestigious British medical journal *Lancet;* the letter went off on July 18. Three days later, concerned that a printers' strike then under way in England might delay publication, Hadlow sent a copy of the letter to Gajdusek at the NIH. The letter was forwarded to Kainantu, where Joe Smadel's near-genius privateer read it in early August.

Hadlow's *Lancet* letter compared the two pathologies. "Each disease," the veterinarian wrote, "is endemic in certain confined populations, whether this be flock or tribe, in which the usual incidence is low, about one to two percent. Sheep, or persons, may become clinically affected months after they have been moved from flocks, or communities. . . . Moreover, scrapie may appear in a previously non-affected flock following introduction of a ram or ewe from a flock in which the disease is known to exist. Likewise, kuru may be introduced through marriage to populations previously free of the disease. Yet . . . the exact mechanism underlying the 'spread' of scrapie or kuru is obscure." So was the cause: "No microbiological agent has been isolated, nor is there convincing evidence that either scrapie or kuru is an infectious disease in the generally recognized sense." The natural history and the signs and symptoms of both diseases were "strikingly similar," Hadlow thought. "Onset of each disorder is insidious and occurs in the absence of signs of antecedent illness." Both diseases developed without fever and were relentlessly progressive. "Both diseases usually end fatally within three to six months after onset." Both had similar signs: loss of coordination "which becomes progressively more severe, tremors, and changes in behavior are features of both diseases." Remarkably similar were the changes in the brain: widespread nerve degeneration; astrogliosis; lack of inflammation; most of all those characteristic holes: "Large single or multilocular 'soap bubble' vacuoles in the [bodies] of nerve cells have long been regarded as a characteristic finding in scrapie; this extremely unusual change, apparently seldom seen in human neuropathologic material, also occurs in kuru."

Then Hadlow came to his electrifying point. Scrapie could be transmitted experimentally from sheep to sheep and sheep to goat. If a disease of sheep in Britain and a disease of humans in New Guinea were improbably alike in so many other ways, then it might follow that kuru could *also* be transmitted experimentally—thus proving that it was also an infectious disease. Hadlow didn't say what was obvious, that scientists couldn't experiment on humans. He proposed the next-best alternative (from the human point of view), that "it might be profitable . . . to examine the possibility of the experimental induction of kuru in a laboratory primate." Successfully infect a chimpanzee or a monkey with kuru and it would then be possible to pursue laboratory studies to identify the infectious agent.

Hadlow's letter must have stunned Carleton Gajdusek. Despite his encyclopedic knowledge of human disease, Gajdusek was a city boy, and he had never heard of scrapie before. He wrote Hadlow on August 6 that he was "deeply grateful" for the veterinarian's letter, but he chose not to admit his ignorance of scrapie. He also chose to give the impression that experiments of the sort Hadlow was proposing had already begun. They had "thus far had poor luck with inoculation experiments," he wrote, "and the possibility of doing more . . . has, until now, been small. We are, however, proceeding accordingly at the present time and frozen and fresh materials are being injected into a number of animal hosts during this year's work on kuru." Which seriously shaded the truth, as Gajdusek admitted at a conference several years later. Before the arrival of Hadlow's letter, he said on that occasion, "infection had . . . seemed a very unlikely . . . possibility for kuru." The letter had "forced [him] to reconsider the problem." Good scientists are competitive, and none more so in those early days of kuru research than Carleton Gajdusek. There had been modest efforts to infect tissue cultures, chick embryos, mice, guinea pigs, rabbits and rats with kuru, and thought of trying primates, but in August 1959 the first chimpanzee inoculation—deep in a government wildlife preserve outside Washington, D.C.—was nearly four years away.

Third Connection . . .

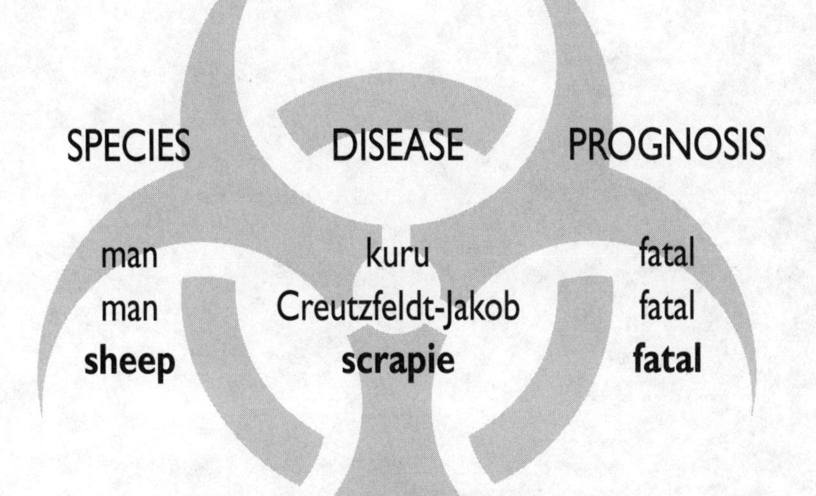

SPECIES	DISEASE	PROGNOSIS
man	kuru	fatal
man	Creutzfeldt-Jakob	fatal
sheep	**scrapie**	**fatal**

FOUR
Across the Species Barrier

New Guinea Eastern Highlands; Bethesda, Maryland; Hamilton, Montana, 1959–1963

CARLETON GAJDUSEK left New Guinea at the beginning of September 1959 to return to Bethesda to find out more about scrapie. Bill Hadlow's realization that scrapie and kuru might be connected opened up promising new territory in kuru research. If kuru and Creutzfeldt-Jakob disease were uncommon conditions of unknown origin, the sheep disease had a long history and had been proven to be infectious.

Joe Smadel knew about scrapie. Gajdusek learned that there had been research on scrapie ongoing at Compton and at the Moredun Institute in Edinburgh, Scotland, since before the Second World War. He learned that a pathologist in Iceland named Björn Sigurdsson had described an Icelandic form of scrapie, called *rida,* in 1954, and had proposed that such unusual infectious agents should be moved into a new class of their own and called "slow viruses." Our usual idea of infectious disease, Sigurdsson had pointed out, came from acute infections such as measles, polio and flu, where a microbe enters the body, establishes itself, multiplies and spreads. Within a few hours or a few days, as a result, symptoms appear, the body takes defensive measures and "a period of struggle between the invader and the host follows lasting usually for a few days," after which the host either dies or recovers. Besides acute infections, Sigurdsson had noted, there were chronic infections such as tuberculosis and malaria, where the invader and the host struggled back and forth sometimes for years in "a long and dubious battle."

With *rida*—scrapie—the battle was long, but it wasn't dubious. The disease followed a more-or-less standardized course just as regular as the course of the acute infections, but the incubation period was extremely protracted. An even more important difference, Sigurdsson felt, was the lack of an immune response—of inflammation. "One is even sometimes tempted to suspect that there is no effective immunity response in the slow infections, which seem to progress unhampered for long periods of time until they kill. If an immunity develops at all it must be a very ineffectual one." Since the immune response represents the body's effort to destroy the foreign protein of the infectious agent, Sigurdsson wondered if the infectious agents of the slow viruses might be so well adapted to their hosts, "so well camouflaged," that they looked to the body like its own tissue. As with amyloid plaques, that conjecture would prove to be another important clue, but neither Sigurdsson, Gajdusek nor anyone else knew how to follow it to its conclusion at the time.

Gajdusek still personally thought heredity would turn out to be the more likely explanation for kuru, but he accepted the necessity of ruling out a slow virus as a cause. If kuru proved to be infectious, the slow-virus model would explain why the small laboratory animals his colleagues had inoculated had not incubated the disease. The animals had been observed for only a few months at most, while a slow virus's incubation period might well be measured in years—perhaps even longer than the animals' life spans. Hadlow's suggestion that kuru should be inoculated into nonhuman primates made sense immunologically, because chimpanzees and monkeys were man's closest animal relatives and therefore the most likely to be susceptible to the disease. But to see results,

Gajdusek now realized unhappily, he would have to be prepared to support the inoculation experiments for years.

Hadlow and Gajdusek met for the first time in Washington at the end of November 1959. The scrapie eradication program that the United States Department of Agriculture had instituted was infuriating sheep farmers. The federal government paid compensation for the wholesale slaughter of scrapie-infected flocks, but the payments were too low to cover the cost of replacement of high-quality purebred animals. Nor did sheep breeders want to destroy purebred lines that they had spent years improving. They found support for their dissent in the work of an influential but misguided Oxford University researcher named H. B. Parry, who was convinced that scrapie was hereditary despite repeated experimental transmissions from sheep to sheep and from sheep to goats. The veterinarian in charge of the USDA scrapie eradication program, James Hourrigan, decided to put together a dog-and-pony show to present the facts to the irate U.S. sheepmen. Hourrigan invited Hadlow, Compton director William Gordon, Moredun Institute director John Stamp and another USDA man to form a team. They started with presentations in Washington on November 23. Hadlow remembers noticing Gajdusek, "the young man with the crew cut who stood silently in the back of the room while I gave my talk," and meeting with him afterward. The two scientists compared notes on kuru and scrapie and discussed inoculating primates.

The Hourrigan scrapie tour faced down hostile sheep farmers in Columbus, Ohio, Chicago, Denver and San Francisco. "It didn't come to any punch-up," Hadlow told me, "but it got quite vocal and a little rowdy. Jim handled it well." Gajdusek negotiated a new visa from the Australians and returned to New Guinea early in the new year, 1960, to find Vincent Zigas drinking and dithering in the throes of a domestic crisis—Zigas's wife had declared herself sick of kuru and had decamped for Sydney. Another grueling round of bush patrols extended into May, during one of which Gajdusek barely escaped attack by hostile warriors of a linguistic group known around New Guinea in those days as the Kukukuku, a swear word meaning something like "motherfuckers." The Kuks (known today as the Anga, a name Gajdusek coined) were the Apaches of New Guinea and possibly the island's true aboriginals, with archeological traces dating back sixty thousand years—ferocious, arrogant and, predictably, Gajdusek's favorites. They were the Fore's immediate neighbors across a powerful river, the Lamari, but despite their proximity to the Fore and their similar material culture and diet, they showed no signs of kuru. They served Gajdusek as a control group for epidemiological comparisons and he studied them extensively. "Except for a very few days of this year," he wrote Joe Smadel that May, "I have been living and working in villages of so-called 'savages' and in many cases in villages not previously visited by civilization and never sojourned in by civilized man." He went on to report the results of his studies of kuru, leprosy and other diseases, including his collection of more than a thousand blood and urine specimens and "many further autopsy specimens . . . still 'hardening' in fixative at Kainantu." Personally more meaningful was his extended encounter with these premodern peoples. "I cannot overemphasize that all this is of minor importance in my mind to the observations and experience I have gained in close association with these oddities and 'orchids' of human culture—fragile and soon to be extinct, but revealing man and his condition as it may never be seen again!"

Gajdusek was disturbed to be an agent of disturbance. He wrote in his journal that he liked primitive cultures better than his own. He was a doctor, however, committed to healing the sick:

> It is a dreadful anthropological shame, and a loss to humanity of one of the most colorful cultures to decorate it, but these orchids of culture cannot survive on the artificial nurturing soil of civilization. . . . We cannot cure their yaws and ulcers, save their dying children, remove their arrows and treat their wounds without coming to them. We cannot come to them without bringing ourselves and our life into their horizon and to then refuse their request to see the outer worlds, or agree with those who would come and study them, observe them, and especially those who want to "help" or change them in any way (including to stop warfare, murder, fear, superstition, famine or pestilence) and who would yet "leave them as they were, primitive and picturesque". . . is an insult to their human aspirations and intelligence and will never do. By coming we commit ourselves to the change and are agents of it. The change disturbs us for we know better than they do how pallid and barren and how unsatisfying the fruits of civilization can be at times.

His personal resolution of the conflict was simply to spend as much time in the Eastern Highlands as possible, living, studying, experiencing and recording New Guinea life in his journals and on film. At some point he observed a cannibal feast, he says, and filmed it, but he has never released the film. "Watching children trained in warfare," he told Smadel in his May letter, "babies crawling over corpses, youngsters undergoing nose piercing, isolation, and other 'initiations,' and patterned homosexuality, juvenile heterosexuality, polygamy, and many other psychosexual patterns in action, the nonsense of much of our pseudo-scientific speculations and philosophizing in medical journals is immediately apparent. Just what to do with these observations, how to make and record them and what to conclude, however, are no simple problems and I will not be quick in solving them."

Returning that winter to the U.S. (after a side trip across the Libyan desert investigating malaria), Gajdusek stopped off to visit the major scrapie research centers at Compton, Edinburgh and Iceland. "I came back," he wrote later, "with the conviction that we had to pursue urgently such inoculations and long-term observations of animals, especially primates." In London he recruited an experienced neuropathologist named Elisabeth Beck to study kuru brains; Beck had dissected a series of thirty-four scrapie brains for Parry at Oxford and knew the characteristic lesions of scrapie cold. Gajdusek arrived in Bethesda with a pocketful of sealed test tubes of scrapie inoculant from Compton in violation of the USDA quarantine, avoiding bureaucratic delays but privateering himself into serious trouble later. Scrapie would give the team he planned to put together for primate research experience with slow-virus studies, starting with an attempt to transmit the sheep disease to mice.

Gajdusek had more in mind than scrapie and kuru. With characteristic bravado, he set his sights on the whole loathsome brood of chronic human neurological diseases, including Parkinson's, ALS and multiple sclerosis. Doctors had long suspected that these diseases originated in infection, though no disease agents had ever been demonstrated. Organizing a lengthy primate study would make it possible to mount transmission experiments for a range of such diseases and help justify the major expense.

Scrapie's long incubation period—five years in sheep, a full year even in goats—appalled Gajdusek, who was a virologist of acute infections by training. If kuru required as long to incubate as scrapie—and it might need even longer—then experiments carried out in the usual way, serially, might take forever. Infect animal A, wait until it showed full-blown symptoms, sacrifice it, demonstrate the disease in its brain, use that brain tissue to infect animal B, wait until *its* full symptoms appeared—what young researcher could afford to devote ten years or more of his career to one or two animal passages of a disease agent that might or might not be infectious? Mice had recently been successfully infected for the first time with scrapie at Compton, and they might passage the disease more quickly (it was to get that work started that Gajdusek had illegally carried scrapie inoculant across the USDA's quarantine). He also realized he could mitigate the incubation delays by running some experiments in parallel, injecting a number of primates simultaneously with inoculants from different kuru cases and with different dilutions of material from the same case.

But whatever biological legerdemain he applied to the problem, he would still be stuck with organizing a large operation that he didn't really believe in and then waiting years for results. He decided he would have to sell the project to what he calls "more patient and less ambitious colleagues" and then "await their first success with skepticism" before entering the field himself.

He needed a location, he says, "where nobody could see what we were doing." The federal government commands vast resources; Gajdusek found his hideout in a facility operated by the Rare and Diminishing Species Program of the Department of the Interior's Fish and Wildlife Service: forty thousand acres of restricted lakes, marshes and woodland between Washington and Baltimore, south of Fort Meade, a place called the Patuxent Wildlife Research Center. Deep inside Patuxent, the remains of a farm lot—a big cattle barn and a smaller dairy barn set at right angles to each other—gave him a nucleus of buildings. The Fish and Wildlife Service agreed to the project in 1961, but negotiations between the NIH and the Department of the Interior, as complicated as an international treaty, would drag on for two more years before construction could begin on an uninsulated cinderblock building to house the primates.

In summer 1961, Gajdusek and Joe Smadel tried to recruit

Bill Hadlow to run the primate inoculation program. By then, Hadlow had returned to Montana and was preparing to do scrapie research at the Rocky Mountain Laboratory. "They told me they were going to inoculate chimps and do what I told them to do," he recalls. He concluded unfairly, he says, "that anyone who took the job would become little more than an exalted handler of apes." So he responded, no, he wasn't interested. "Hell, I'd just gotten settled in Montana and my wife just had the baby and I didn't feel like moving again." He returned without regrets to Montana and scrapie research.

Gajdusek and Smadel next focused on a short, blunt, determined native Washingtonian Ph.D. virologist named Clarence Joseph Gibbs, Jr., who specialized in insect-borne viruses and had already developed a successful vaccine for Rift Valley fever, a disease that affects both animals and humans in East Africa. Joe Gibbs had worked for Smadel at Walter Reed. He stopped by Smadel's office one day to let the NIH associate director know that he was planning to take up a Rockefeller Foundation fellowship for research in Brazil. "Smadel's reaction was immediate and violent," he remembers, "and in his inimitable fashion he pointed his finger in my face and said, 'Goddamn it, Gibbs, you're not going to Brazil!'" Gibbs asked Smadel just where he could expect to be banished. Smadel told him he was going to Patuxent.

Gibbs was reluctant to leave his arboviruses for the unknowns of kuru. Gajdusek courted him over lunch, promising him that he could continue his own work, not saying that it might help keep him busy while he waited interminably for kuru to incubate. "We can pick and choose what we want to do," Gajdusek told him. Gibbs talked to Bill Hadlow as well. It was Smadel who finally won him over. "My decision was a simple one when Dr. Smadel told me in very fatherly terms, 'Joe, I want you to undertake this work and I guarantee that within five years you will either have golden positive or golden negative results and I am sure they will be golden positive.'" Gibbs signed on for what Gajdusek now identified as the "Study of Slow, Latent and Temperate Virus Infections"; in the months to come, the sturdy young virologist would inoculate thousands of mice with scrapie, kuru and other human neurological diseases. With Gibbs installed at the NIH to organize the transmission studies, Gajdusek returned to New Guinea.

He found the Fore in crisis and a husband-and-wife team of anthropologists taking up residence in the South Fore village of Wanitabe. Robert and Shirley Glasse (now Shirley Lindenbaum), an American and an Australian, had encountered the village on a Land Rover reconnaissance southward from Okapa. The village big men had welcomed them and offered to build them a house. They accepted, paying for the thatched construction with salt, tobacco, cloth, beads and other trade goods. Later the Glasses learned that the big men had been cheated a decade earlier by North Fore con men and had interpreted the anthropologists' arrival as vindication. The North Fore had sold the Wanitabe big men a phony cargo cult, instructing them in transforming local raw materials magically into Western trade goods in exchange for pig meat. The big men told the Glasses they had loaded sand, pebbles and wood slivers into a special "cargo" house—the North Fore had warned them not to peek while the magic worked—had "smoked pipes and eaten [hallucinogenic] bark until our throats were parched and our heads spun" but found no knives, guns, bullets or cowrie-shell money when they finally entered the house to inspect. They proceeded to work the same con on their neighbors farther south, recouping some of their losses, but their gullibility had thoroughly embarrassed them until the Glasses arrived. "Our talk of building a house and settling in the community," Shirley Lindenbaum writes, "was viewed as the delayed arrival of the 'European' goods they had expected a decade earlier." The Wanitabe big men took to calling the house they built for the Glasses the village "store."

When she began working among the South Fore that summer, Shirley Lindenbaum was twenty-eight years old, a warm, open, attractive Australian with a rose-petal complexion and dark blond hair. She had been an English major at university in Australia, reading Old Norse sagas, when her sister-in-law returned from studying anthropology at Oxford and the two spent a day together at the beach. The intellectual power and the philosophic reach of anthropology, as her sister-in-law communicated it, had enthralled her. The Australian feeling of living at the far end of nowhere conditioned her for such a revelation. Down under, at the bottom of the world, you thought you had to go *up* to get to European civilization, with Asia in the way. Lindenbaum's father was a timber merchant. "My family cut down the primeval forest of eastern Australia, I'm afraid," she told me ironically. Her generation wasn't the pioneer generation, more the offspring of cities and suburbs, but the yearning was still there. After her encounter with her sister-in-law she moved to Sydney, sold her Volkswagen—her prize possession—used the proceeds to pay for graduate study in anthropology and never looked back.

The Fore, she discovered, were "very charming people even though gender relations were difficult." Easier for her, coming in from the outside. "European women are transitional genders in that setting because we act like men and have wealth like men. Though you wouldn't want to step over a baking sweet potato, which I've done, or the men would have to throw it out—you might have contaminated it with a drop of menstrual blood." The Fore women welcomed her. "They really liked having another woman there. They felt my body to see if I had body parts like theirs. I hung around with them in ways that mission women didn't—the mission there at the time had a huge fence around it and the missionary's wife and daughter didn't come out. So the Fore didn't actually know any European women. I loved it. I never felt afraid and I was surrounded by warm people."

Lindenbaum interviewed the women about their lives. Eventually, they recalled their cannibalism for her. One woman remembered watching, feeling scared, while her older sister partook. Eating the dead was never routine. "They hid in the garden, they did it at night, did it away from the eyes of people, they did it in the old sugarcane gardens and sometimes even in the burial ground. It stirred erotic excitement and maybe gender power. Women were attacking men's bodies. It was their own domain and men were kept out."

The years of the Glasses' fieldwork—mid-1961 to mid-1963—coincided with the height of the kuru epidemic in the South Fore. Annual deaths from kuru had peaked in 1959 at more than 130 but rose again above 120 in 1963, almost all of them women and girls. "Of 125 Wanitabe males over the age of 21," Lindenbaum writes, ". . . 63 had no living wife, 10 had never married. That is, just under half the adult Wanitabe men were without wives in 1962. There is scarcely a man in South Fore who has not lost a number of immediate kin to kuru. One Wanitabe big man has seen the death of his mother, his half-sister, four wives, one son and one stepson. Pigo of Wanitabe, a man about 45 years in age, had three wives and three daughters die of kuru, while Anatu of Kamila has been deprived of his mother, four wives, one sister, one daughter and one son. . . . The genealogical records convey a sense of obliteration."

In response to their increasing fears of extinction, the Fore called great public meetings where men and women both appealed for an end to Fore sorcery. At one meeting, Lindenbaum heard a woman say:

> Why are you men killing off all the women, stealing our feces from the latrines to [make sorcery bundles and] perform sorcery? We women give birth to you men. Try to find one man who is pregnant now and show him to us. Or go and search the old burial grounds and bring us the skull or bones of one man we women have killed. You won't be able to find any. You men are trying to wipe us out.

Since ordinary sickness would take men and infants as well, the fact that most kuru victims were female proved to the Fore that sorcery was responsible.

Besides attending public meetings, Fore men also hunted down men they believed to be sorcerers and killed them in reprisal. The hunters used a specialized attack called *tukabu* against sorcerers: they ruptured their kidneys, crushed their genitals and broke their big thigh bones with stone axes, bit into their necks and tore out their tracheas, jammed bamboo splinters into their veins to bleed them. Gajdusek saw more than one mutilated corpse and treated a man severely wounded in a *tukabu* attack who had barely escaped with his life. As a result of the kuru epidemic, men outnumbered women by an average of two to one throughout the South Fore; in some villages the ratio reached three to one. The loss of men to *tukabu* killing reduced the disparity, as if some hardwired biological balance had to be struck.

Fore men moved reluctantly to take up what had formerly been women's work, hauling firewood, tending gardens, cooking food and feeding their children. Yet the Fore were the first to notice that the epidemic pattern was changing. Deaths were declining in the North Fore, if still high in the South. The most encouraging change was that fewer children were dying of kuru. Forty-six had died in 1957 and forty-two in 1959, but 1961 saw only fourteen child deaths.

At the end of 1961, Michael Alpers arrived in the South

Fore from Australia with his wife and ten-month-old daughter, assigned for two years as a government physician to study kuru. Gajdusek described him in a journal entry at the time as "young, quiet and conscientious." Alpers had heard tales of the wild American, but despite his insertion into the scene to maintain an Australian presence, which the wild American might have resented, he and Gajdusek got along well. On a patrol together in the spring, they discovered an unmapped New Guinea volcano. "We walked this large crater," Alpers recalls, "and we found a yellow lake bubbling in the middle of it. Every step we took there was hydrogen sulfide coming up. We got halfway around this thing and stopped, Carleton and I looked at each other and we said, 'Fuck, this is a volcano.' Took a while for the penny to drop." Alpers remembers the patrol because a radio message came through for Gajdusek that someone back home was dying. Gajdusek didn't know whether it was his poet brother Robin, Joe Gibbs or Joe Smadel. He cut the patrol short. It was Joe Smadel, mortally ill, and in July, reluctantly preparing to leave again for home, Gajdusek confided to his journal his apprehension at returning to "civilization":

> I have lived in a world of children and child humor, child fantasy and child passions for four decades and I have not had to occupy the irrevocable role of a father to do so. If only I can grow old, foolishly old, in this same world . . . I shall be most fortunate! I leave now the most playful and rewarding of all child worlds, and already I regret it! . . . These fascinating villages mean more to me than Washington and Boston. Frankly, I am a bit afraid to come home and meet mother, meet Joe Smadel dying—and a bit reluctant to get "involved" again with family and friends and their complex lives.

• • •

At the Rocky Mountain Laboratory, Bill Hadlow developed a sideline studying mink. The fierce little predators, prized for their fur, are cage-raised on ranches in the northern United States, in Canada and across northern Europe and Asia. Hadlow was investigating a severe infection of mink, Aleutian disease, that caused millions of dollars of losses annually. "Mink was a new animal for me," he recalls. "I had to learn to work with them. I designed mink sheds, designed cages, found a source of feed, found a source of mink just twelve miles down the valley on a small mink ranch. We had an arrangement with the rancher that he wouldn't bring in any mink from outside, he wouldn't take his mink to shows, he wouldn't let in other mink ranchers. I would postmortem everything that died. That worked out fine. People at other laboratories would bring in mink for experiments and half of them would come down with a disease before they could even get started. I always had clean mink."

Early in the summer of 1963, Hadlow took a call from the head veterinarian of the U.S. Public Health Service asking him to examine diseased mink from a ranch in Blackfoot, Idaho. The veterinarian told Hadlow the ranch was suspected of harboring toxoplasmosis, a parasitic disease that affects the central nervous system of warm-blooded animals and that can infect humans. "I'd been fiddling with a little toxoplasmosis here at the lab," Hadlow remembers, "and I said, all right, tell him to bring up a mink. So he brought up a couple of mink on a Saturday afternoon and they were showing good neurologic signs. We killed them, did a postmortem and sectioned the brains. Ha, looked like scrapie."

By then it was suspected that sheep passed scrapie among themselves by eating afterbirth or picked it up from pasture where other sheep had grazed. But mink weren't known to harbor scrapie. Having ruled out toxoplasmosis, Hadlow discussed his findings with his colleagues. He learned that a scrapie-like disease had decimated mink in northern Wisconsin in 1947 and 1961 and that two mink ranches there had just reported new outbreaks. "That's all I needed," Hadlow told me. "I got there right away." At the ranch Hadlow visited he found confused, dirty mink aimlessly circling their cages with dragging hindquarters and tails oddly arched over their backs. He euthanized ten of the sick animals, lined up the cadavers on a picnic table and conducted rough-and-ready autopsies, sealing tissue samples in glass tubes with a borrowed welding torch, freezing those, preserving the brains in jars, carrying everything back to Montana under his airplane seat. Like the Blackfoot mink, the brains from Wisconsin showed spongiform change, nerve-cell loss and the familiar brown star fields of astrogliosis. Hadlow used the frozen tissue he'd collected to inoculate some of his clean laboratory mink. Eight months later, the inoculated mink began dragging their hindquarters and circling their cages with their tails over their backs.

The disease would be named transmissible mink encephalopathy (TME). *Enkephalos* was the classical Greek word for brain; "-pathy," meaning disease, derives from another Greek word, *pathos,* which originally meant suffering. TME was unknown in wild mink; it appeared to be a disease exclusively of mink raised on ranches. That fact led veterinarians investigating TME to suspect the animals' rations. Mink are carnivores; ranchers typically feed them raw meat, packing-plant by-products, fish, liver and cereal. Packing-plant by-products—waste left over from slaughtering food animals—might include sheep meat or offal, so scrapie was the likeliest agent of TME infection. But the first U.S. outbreak of TME, in Wisconsin in 1947, preceded the arrival of scrapie in the state—the first U.S. scrapie outbreak, also in 1947, had been confined to a farm in Michigan. Which at least circumstantially ruled out scrapie in the 1947 TME outbreak. The veterinarians who studied that first TME outbreak noted in particular that meat from so-called downer cattle, meaning cattle found dead or paralyzed and therefore not considered suitable for human consumption, had been fed to the mink that subsequently died of TME. The mink on the ranch Hadlow visited in 1963 had also been fed downer cattle meat. Yet there was no known disease of cattle that resembled scrapie.

Here, then, was another spongiform encephalopathy of mysterious origin and devastating effect. Twelve hundred mink died in the 1947 outbreak. The outbreaks in Wisconsin in 1961 were less lethal, accounting for twenty to thirty percent of the affected mink herds. The 1963 outbreak Hadlow investigated took another twelve hundred animals—an adult herd mortality of one hundred percent. What the cause might be, no one knew with certainty. Veterinary scientists set to work to see if they could passage TME to other animals and scrapie to mink. Their experiments paralleled Gajdusek and Gibbs's experimental efforts to transmit kuru to chimpanzees—work that in 1963 finally got under way at Patuxent, with astonishing results.

Fourth Connection . . .

SPECIES	DISEASE	PROGNOSIS
man	kuru	fatal
man	Creutzfeldt-Jakob	fatal
sheep	scrapie	fatal
mink	**transmissible mink encephalopathy**	**fatal**

FIVE
The Life and Death of Georgette

Patuxent Wildlife Research Center and New Guinea Eastern Highlands, 1963–1965

JOE GIBBS came to hate using chimpanzees for medical experiments. The bright, engaging primates were too human. Gibbs gave up inoculating chimpanzees with lethal diseases years ago—"I just couldn't do it anymore," he says today. But in 1963, with an epidemic of kuru killing hundreds of Fore women and children and decimating Fore families, everyone involved with kuru research endorsed chimpanzee transmission experiments. The closest available model of the disease process they were studying was scrapie transmission in sheep. Some breeds of sheep were more susceptible to scrapie than others. By analogy, some nonhuman primates would probably be more susceptible to kuru than others if kuru proved transmissible, and the chimpanzee, the primate closest to man genetically, was the likeliest candidate. Carleton Gajdusek and Michael Alpers would supply fresh kuru brains from their outposts in New Guinea. Joe Gibbs in the U.S. would carry the burden of running the experiments.

Gibbs still resides in the family house on Capitol Hill that his great-grandparents built. His father had been a field veterinarian for the State of Maryland and Joe had helped take and save the lives of animals even in childhood. "In the summertime, we used to go around shooting dogs and cats because of the rabies problem," he told me. "My father was the first veterinarian in the United States to initiate free rabies vaccination of household pets. That was back in 1939. It led to a dramatic drop in the cases of rabies in dogs and cats in the following year. He continued that free rabies clinic right up until he died in 1971." Veterinary work prepared Gibbs to do science, and after a stint in the Navy at the end of the Second World War he earned a B.A., M.A. and Ph.D. at Catholic University in Washington—finishing up his Ph.D. part-time at night after he began research at Walter Reed.

When he joined Gajdusek's slow-virus study, Gibbs found few resources and no animals at hand except mice. The abandoned barns at Patuxent needed repair, the cinder-block primate house was still on the drawing board and he had to scavenge equipment. "I was given enough money to buy a single centrifuge and a few reagents," he says dryly. "I furnished the rest of the operation out of the military surplus bin. I started off with two technicians. They'd complain we needed this, we needed that. My philosophy was, we don't ask for anything. We get results and then we won't have to ask."

Chimpanzees in those days cost $300 (compared to $25,000 today). In the first half of 1963, Gibbs acquired three two-year-olds. Gajdusek was in the U.S. at the time and he and Gibbs inoculated the first chimpanzee, A-1, "affectionately named Daisy," on February 17. The third, A-4, which weighed thirteen pounds on its arrival at the NIH during the summer, they held in quarantine long enough to test for tuberculosis and inoculated on September 17. They named A-4 George. The brain tissue used in George's inoculation came from a Fore child, a boy named Eiro. Mike Alpers had done the limited postmortem, taking only the brain, snipping small samples from the fresh brain wearing sterile gloves, placing the samples in sterile bottles, freezing the bottles overnight in the Okapa hospital freezer, personally carrying them by air charter to Lae on the northeast coast of New Guinea, packing them in ice, shipping them on to a laboratory in Melbourne that kept them frozen until they could be dispatched in dry ice to the NIH. ("Then I went back to the patient's village and took part in the funeral ceremonies," Alpers writes, "contributing something to the mortuary feast and mourning as earnestly as the other participants.")

In Gajdusek's lab at the NIH on September 17, Gibbs thawed the brain samples, homogenized them with a blender and diluted the frothy pink homogenate with sterile saline. The two scientists anesthetized George to a drowsy state with ether, drilled a small hole with a dental drill into and through the front of George's skull just to the left of the midline, drew two-tenths of a milliliter of the kuru-infected solution into a syringe and injected the solution directly into the cerebrum. "Within five minutes after inoculation," Gibbs recorded in his clinical notes, "the animal had recovered from the effects of the ether anesthesia and during the ensuing days . . . showed no observable external signs of discomfort." When A-4 matured, Gibbs discovered that he had mistaken its gender. George became Georgette, a little female with jug ears and lively black eyes.

Making do, Gibbs commandeered an unused laboratory building at Patuxent to expand the mouse scrapie study that Gajdusek had begun. He contracted with a private biomedical research facility in Falls Church, Virginia, to house several hundred rhesus and African green monkeys which he and his two technicians proceeded to inoculate with kuru, Parkinson's, ALS and other human neurological diseases. Tragedy engulfed that early work six months along when the private contractor accidentally introduced a virulent tuberculosis into the facility. "It was a tremendous disaster," Gibbs told me sadly. "We had put so much work into getting the animals, inoculating the animals, following the animals, and then to have tuberculosis break out—it just spread like wildfire. There was nothing to do but destroy the whole colony. We did it ourselves, bled them all out, because we wanted their brains anyhow. We saved every brain of every animal and studied those brains." But six months wasn't long enough to incubate kuru.

After the Falls Church disaster, Gibbs started over at Patuxent. By then the cinder-block building was finished—an uninsulated, whitewashed one-story barrack structure with caged outdoor runs along its east side and a propane bottle for heating. In April 1963, Gibbs transferred his scrapie-infected mice from the NIH campus in Bethesda to the vacant laboratory at Patuxent. He needed additional strains of scrapie from Iceland to expand his scrapie studies, so he applied to the USDA for an import permit. That precipitated another disaster, this one bureaucratic. "The guy in charge of the Animal Inspection and Quarantine Division—his name was Dr. Robert C. Reisinger—came up and inspected and he was absolutely astounded. He wrote a letter to the director of the NIH saying we had inadequate containment, inadequate airflow and we weren't mosquito-proof. He concluded that I shouldn't be permitted to do any work. He came down especially hard on what he called our 'irregular method of acquiring scrapie virus' "—Gajdusek's shirt-pocket importations from Compton, coming back to haunt them.

"The next thing I heard," Gibbs continues, "Reisinger was putting together a committee to review the biosafety aspects of this place. Unfortunately for him, he put together a committee of my old buddies and they were all very commendatory. He had a point, though. Officially, we were a totally inadequate facility. His fear was that some of these infected mice would get out of their cages and into the wild mouse population in that area." Gibbs blunted the inspector's assault by gradually moving his ten thousand mice into the basement of the dairy barn across the lot from the primate house. "I had all the cow stalls ripped out and just put new cement on the walls to make it somewhat clean. Put in a couple of window units for air-conditioning and heating."

Reisinger still wasn't happy. He almost succeeded in blocking the slow-virus study. But Gibbs is a tough, determined man; the continued interference was more than he could tolerate. "If I feel I'm right," he told me, "and I feel the other person is way off base, don't get in my way, because I'm going to make sure something happens to resolve the situation. I couldn't get people to understand that these are not contagious diseases, absolutely not contagious diseases. Mosquitoes don't spread scrapie. Neither do mice. I decided Reisinger had to go. I sort of arranged to have him offered a position at the National Cancer Institute and off he went. We ultimately became friends anyhow. He saw the error of his ways."

With his small-animal colonies transferred to Patuxent, Gibbs decided to consolidate his operation. He moved the chimpanzees from the NIH into the primate house, where they joined the collection he'd accumulated by then of seventy-five rhesus, African green, squirrel and cynomolgus monkeys. He added four more chimpanzees—"the building was starting to bulge at the seams." The monkeys stayed caged, but the personable chimps got to play. "I made a point of always going morning and afternoon, late afternoon, to do animal observations. The chimps were still small enough that you could handle them. Play with them. I would get them out, take one by the hands and throw it up into the air and it would do a couple of flips in the air and I'd catch it." They didn't need biohazard containment. "Containment to us meant keeping them indoors," Gibbs quips. One year in the dead of winter the chimps let the monkeys out. The monkeys headed screaming for the trees and then the cold hit them. "You could just go around and pluck them off the branches,"

Gibbs laughs. "The chimps had better sense. They went back inside where it was warm."

Mike Alpers moved to Bethesda early in 1964 to work his way systematically through the voluminous kuru epidemiological records that Gajdusek had accumulated there since 1957. Alpers hoped to find patterns hidden in the records that the team had not yet noticed. He did:

> It was an exciting moment when . . . I added up all the figures by age group and divided them into the early period from 1957 to 1959 and the more recent one from 1961 to 1963. There was an epidemiological impression from the field that the disease was becoming less common in children and this impression was sustained as one worked systematically through the whole file. . . . Suddenly there were the figures in front of us: kuru had declined indeed in children but in the younger age group it had essentially disappeared. This was an exciting new fact about kuru that cried out for an explanation. For the moment, however, all I could do was wonder at this revelation that shone through our dull charts.

The National Institute of Neurological Diseases and Blindness sponsored a working conference on slow, latent and temperate viruses in December 1964. Gibbs and Gajdusek organized the meeting. Most of the major players made their way to Bethesda to stake their claims and report their progress: scrapie specialists from Compton and Edinburgh, neuropathologist Elisabeth Beck from London, Bill Hadlow, USDA scrapie-eradication program director Jim Hourrigan, Gibbs and Gajdusek's NIH colleagues Paul Brown and Igor Klatzo, even Gibbs's bureaucratic nemesis Robert Reisinger. Iceland, still smarting from Gajdusek's invasion of the field, was conspicuously underrepresented, sending only one participant. The discussions covered not only kuru and scrapie but also Aleutian disease in mink, various kinds of encephalitis, multiple sclerosis and ALS. Curiously, no formal papers reviewed what little was known about Creutzfeldt-Jakob disease, though Klatzo repeated his finding that kuru resembled the obscure condition. Alpers reported his exciting discovery that kuru had disappeared in young children. A Compton researcher asked him if the disappearance coincided with the end of cannibalism among the Fore. Alpers said it did, and pointed out that the transfer of kuru infection by cannibalism had been "Dr. Gajdusek's first hypothesis," but minimized the connection. Gajdusek chimed in: "I do not believe that cannibalism is the answer, but if one wants to consider all aspects of it, the Fore did often eat partially cooked brain which was contaminated with putrefying flesh and viscera."

In the 1880s, the great Robert Koch, the German physician who first identified the organisms that cause anthrax, tuberculosis and cholera, defined the steps medical researchers logically had to follow to prove that a particular disease agent causes a disease. The three principles he defined are known as Koch's postulates:

1. The agent must be present in every case of the disease.
2. Inoculations of pure cultures of the agent must produce disease in animals.
3. Cultures of the agent purified from such diseased animals must produce the disease again and repeatably.

Gajdusek and his colleagues were using Koch's postulates as a guide in their effort to transmit kuru to primates. They had assumed the agent was present in their kuru cases and for purposes of the experiment they had also assumed that it was transmissible. If their animals sickened with a disease that looked like kuru and if brain tissue from those diseased animals transmitted kuru to another generation of animals, then it would be incontrovertible that kuru was infectious. To satisfy Koch's postulates fully they would then have to grow their kuru agent in pure cultures outside the body—"in vitro," as they called it, using the Latin word for glass. Not every known disease could be grown in vitro, however. Scrapie was an obvious exception, which meant scrapie researchers had to rely instead on the techniques developed originally by Louis Pasteur of passaging the scrapie agent through animals. At this point, at least, so did Gajdusek.

Besides reviewing the kuru records in Gajdusek's laboratory, Alpers also helped Gibbs with his primate studies. "I was working as a chimpanzee neurologist," the Australian physician writes, "following the animals that had been inoculated with the kuru brain material so carefully obtained from my patients in . . . New Guinea. . . . I spent many hours patiently and regularly documenting the clinical state of these animals." So did Gibbs; so did Gibbs's tall, reliable lab-technician assistant, Mike Sulima. It was Sulima, on June 28, 1965, who first noticed a change in Georgette's behavior, writing in his daily record: "Does not appear to be as active as usual. Stays by herself in corner." Gibbs recalls that Georgette "was starting to withdraw and had a vacant look in her eyes. And we noticed a little shiver, as though a cold wind had come through and suddenly hit her."

Gibbs had never followed a kuru patient. "I didn't want to see a case of kuru prior to observing these animals, so that my observations wouldn't be influenced by anything I had seen before. We watched this situation develop over time, over weeks. It was clearly a progressive course." Georgette's lower lip started to droop. Sulima noted:

> *14 July 65.* Appears to have the "shakes." Trembles at frequent intervals. No sense of balance. Fell off stoop in

cage. Moving around very slowly. Jaw hangs open constantly.
15 July 65. Tremors continue. Fell off top of cage today.

"We watched how Georgette started to hang on to the side of the cage," Gibbs elaborates, "because she was unstable. Sometimes, getting down off the perch in the cage, she would

Georgette with drooping lip.

fall. We watched how she would suddenly grope for the food that was right there in front of her and how her hand would shake when she would bring the food up."

Alpers made the call: "I remember one day going into the anteroom where I wrote my clinical report and after dashing it off in the usual fashion I made my summary at the end, almost mechanically: 'clinical impression—kuru.' The word leaped from the page." From Patuxent, Alpers drove directly to Bethesda and found Gibbs. "I said, Joe, this animal has got kuru. He just laughed: 'You're imagining it.' And I said, 'We've got to get Carleton in.'" By then, Daisy had also begun showing signs.

Gajdusek was in New Guinea. Gibbs cabled him the news of the two chimps' condition and asked him to fly home to examine them himself. "He flew straight back," says Alpers, "somewhat annoyed at having to interrupt his field trip—he'd been seeing kuru patients." The morning after Gajdusek arrived, July 20, they drove out to see the chimps. "Georgette performed even better than she had done with me," Alpers recalls. "I'd had to sort of drag it out with clinical examination, but she came into the room with Carleton there and she then started walking and stumbling all over the place—behaving just like a kuru patient. Carleton had been seeing kuru patients in the field twenty-four hours beforehand and he couldn't believe it. It was just a magnificent performance on the part of this chimpanzee. So that was very convincing, that was the point where we decided, well, this probably really is it."

Gajdusek still had his doubts. Waiting for an airline connection in Fiji on July 31, he wrote in his journal that it was "more than ever likely that some transmissible agent with long latency might underlie kuru," but that it would be "years before anything can be really proved from chimpanzee inoculations." He went on to rehearse the requirements of Koch's postulates: "Thus, even if this chimp

[Georgette] develops full-blown kuru-like disease from which she dies, and if her brain shows kuru-like pathology, we will have to reproduce the disease by reinoculation of [kuru victim] Eiro's brain material into other chimpanzees and also transmit it in passage to new chimpanzees, and finally do so with filtered material and by various routes of inoculation. All this will take years, at least. . . ." When Gibbs had cabled him about Georgette's condition, Gibbs writes, Gajdusek's first reaction had been to wonder if they had botched the experiment. "His reply was, "Say nothing of this development, we may well have contaminated the animals with scrapie.' I doubted this very much since I had been very careful to keep the scrapie animals and their inocula totally separate from the human disease program."

Either Gajdusek was finding it difficult to switch back to the idea that kuru was infectious rather than hereditary or he was reining in his hopes, probably both. A few weeks later he was writing to Gibbs that he was "as excited as you are" and to Alpers not to wait too long to "plan a prescheduled and thorough sterile-technique autopsy" once the disease seemed to be "progressive and unrelenting—if that is the case. . . ." In the meantime, Gajdusek wrote Alpers, "keep watching it and documenting it exquisitely with cinema and repeated neurological examinations." To the assistant director of the NIH who approved funding for his laboratory now that Joe Smadel was gone, Gajdusek wrote that "to suddenly have such a remarkable cerebellar disease appear in one of our earliest kuru-inoculated chimps—and perhaps also in another—is more than we had any right to expect. To not jump on this lead—the most exciting lead we could ever find—with a vast effort would be foolish and make nonsense of our whole expensive endeavor." If the prestige of discovering a new disease was like the prestige of discovering a new element, the two discoveries also usually garnered similar rewards: each was likely to earn its savant a Nobel Prize. On the other hand, few blunders damage a scientist's reputation more than a mistaken claim—and the larger the claim, the worse the damage. No wonder Gajdusek careened between elation and caution.

Once Georgette's kurulike condition became obvious, she quickly deteriorated. Gibbs's notes chronicle her decline:

> *9 Aug 65*. I spent considerable time observing Georgette today before letting her out of her cage. I quietly took a position where she could not see me but I had a good view of her. I was alarmed to note that the animal, sitting back on the ledge in her cage, was besieged with what I would call "classical shaking chills." . . . The animal was tightly drawn up with arms wrapped about her trunk and with knees and legs drawn up to the body. . . .
> *13 Aug 65*. Friday the 13th. Georgette again had chills; she fell out of her cage—but the fall was nowhere near as hard as many of her other falls; quite lethargic in [her] outside run—tremors continue as does the "droopy lip" condition. Am convinced she has lost depth perception in regards [to] the left hand—reaches for food but *always* grabs my arm up near elbow, missing the food target completely. . . .
> *7 Sept 65*. There appears to be a further downward trend in the animal's condition today. This morning she was on the bottom of her cage and although her respirations were regular, I first thought she had "sudden death" during the night. . . . Definite tremor, erect hair follicles and uncoordinated and weak movement. . . .

"It got to the point," Gibbs told me, "where I had to establish twenty-four-hour-a-day nursing. The two animals—Georgette and Daisy—were absolutely unable to do anything. We brought them up to the main lab and set up a little nursery for them." Gibbs's notes go on:

20 Sept 65. I don't know how much longer this animal can continue to degenerate. She is starting to appear emaciated and grossly weaker. . . . We fed her again by hand. . . .
27 Sept 65. I fed the animal by hand today; managed to get 3 apples and 3 pieces of whole wheat bread soaked in milk and cream into her. . . . Shows a loss of 600 grams since 24 Sept 65. Am quite worried about fluid intake. . . .
7 Oct 65. Georgette remains apathetic—though appetite is good. I am firmly convinced that this animal will never recover. She is getting weaker though the strength in her arms remains very good—she can support her own weight up & down. My life is getting to be hell for fear of sudden death or complicating secondary infection—I think we must sacrifice [her] very soon. . . .

"Finally," Gibbs recalls painfully, "we decided that this had gone on long enough. We flew Mrs. Elisabeth Beck in from England and we set up to sacrifice the animal and do the autopsy. Which was tough on all of us because we'd become so close to these very remarkable animals and would feel their loss." They laid elaborate plans for the autopsy, communicating back and forth with Gajdusek in New Guinea.

On the morning of October 28, 1965, Alpers filmed Georgette's behavior one last time and conducted a final examination:

Attempted to show her poor vision: ability to smell an apple, to root for it, but no fixation of vision—shown by the fact that she appears to want the apple and is searching for it but will not follow it with her eyes. Menace response only occasionally present. Is still able to move all four limbs. Attempts to "walk" or rather drag her body towards an apple offered by placing it in front of her nose as she leans forward to place her head on the ground. . . . Essentially unchanged from previous day.

They said goodbye to Georgette, anesthetized her and painlessly killed her by draining her blood, putting her out of her misery. "It was more of a harvest of the body than an autopsy," Alpers recalls. "Joe and I did it and Elisabeth was there and we put every bit of tissue from the whole body into a whole series of bottles for frozen sections, trace-metal analysis and what have you." A visiting scientist's notes record the harvest: a few grams of brain samples for virus studies, the rest of the brain suspended in formalin for fixation; the entire length of the spinal column together with part of the rib cage and pelvis, also fixed in formalin; the heart, lungs, liver, gallbladder, spleen, pancreas, kidneys, adrenals, left ovary ("right ovary not found"); the entire gastrointestinal tract, lymph nodes, thyroid gland and thymus gland. "The left upper and lower extremities were disarticulated, skinned off and fixed separately in formalin," the notes record. "The remaining carcass of the chimp was likewise fixed in formalin."

Beck was assigned to section Georgette's brain and look for pathological changes. She flew back to London, carrying the brain with her in its sealed jar. "Before she could take any sections," Alpers explains, "she had to wait for three weeks at least for fixing, so we were all sort of waiting and waiting. She's a professional—she wouldn't hurry. Finally, she decided it was ready to go. She sectioned it and started looking and then one morning we got a telex."

The telex as Alpers remembers it was brief and to the point:

PATHOLOGY OF GEORGETTE INDISTINGUISHABLE FROM HUMAN KURU.

Fifth Connection . . .

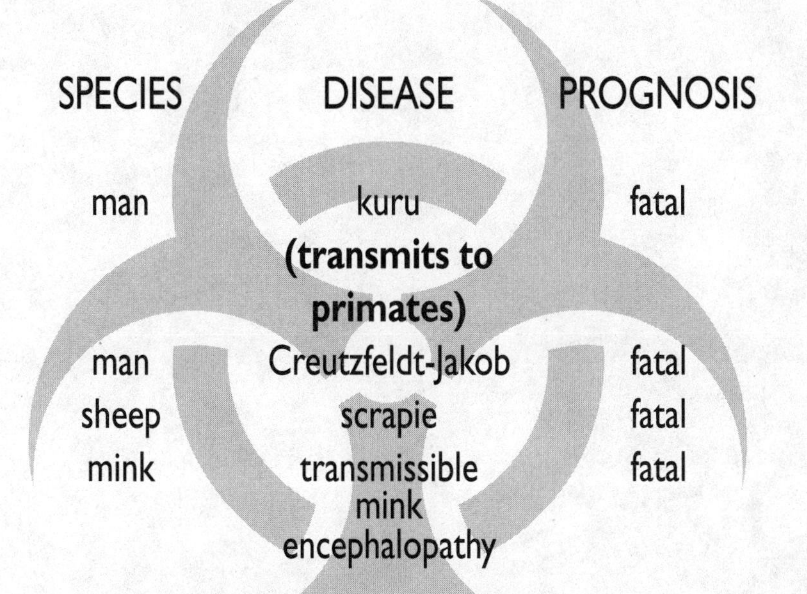

SPECIES	DISEASE	PROGNOSIS
man	kuru **(transmits to primates)**	fatal
man	Creutzfeldt-Jakob	fatal
sheep	scrapie	fatal
mink	transmissible mink encephalopathy	fatal

SIX
The Cannibal Connection

New Guinea Eastern Highlands, 1962–1963 / New York and Bethesda, 1966–1968

SHIRLEY LINDENBAUM REMEMBERS VIVIDLY when she and her anthropologist first husband, Robert Glasse, grasped the connection between kuru and Fore cannibalism. "When anthropologists go into the field," she writes in retrospect, "they leave family, friends and entertainment, but often take pieces of their own culture with them. . . . In our case, before we left Australia [for New Guinea] in June 1961, we arranged to have a wide range of serious and not-serious reading matter sent to us." One periodical the Glasses subscribed to was *Time*. It arrived a month or more late, but it gave them a view of events outside the limited world of the Fore. In June or July 1962, an issue of *Time* arrived in Wanitabe that contained a curious story. A scientist, *Time* reported, had trained planaria—small flatworms that live in water and moist soil—to find their way through a simple maze, had chopped up these educated specimens and fed them to other planaria and had then demonstrated that the cannibal planaria "remembered" the maze.* "The idea of the transfer through cannibalism of some key agent," Linden-

---

*The work was later discredited.

baum explains, "provided us with a 'model' for considering the parallels in our own data on Fore cannibalism."

The Glasses didn't act on their insight immediately. Unlike laboratory scientists, anthropologists aren't often in competition to publish their results, because they're usually the only workers in their field of science who are observing a particular cultural group. Carleton Gajdusek kept voluminous journals of his experiences in New Guinea, and Mike Alpers incorporated anthropological observation into his epidemiological studies, but both men were physicians, not professional anthropologists. After *Time* alerted them to cannibal planaria, the Glasses continued to collect data on Fore cannibalism and on kuru—but now they had a clear sense that the two might be connected.

Even before the Glasses had arrived in New Guinea, two American anthropologists at Tulane University, Ann and J. L. Fischer, had armchaired a connection between kuru and cannibalism by working their way through the findings of a team of anthropologists who had studied the Fore in the early 1950s, Ronald and Catherine Berndt, as well as the many papers on kuru that Gajdusek, Zigas and various Australian investigators had published. The Fischers had summarized their findings in an American journal in spring 1961:

> The Fore habit of eating corpses suggests a way in which a viral agent might be passed. (A toxic one might also be passed in this way.) Victims of some kinds of sorcery are not eaten by the Fore, who fear they might be poisoned, but kuru victims are evidently not included in this category. . . . Berndt says that it would be difficult to prove, but that women may eat more human flesh among the Fore. In any case, corpses are said to be consumed in all stages of decay and with all degrees of cooking. If this is the case, women are probably more likely to eat raw corpse than men are.

> An infectious agent could explain why kuru seems to be passed down through maternal rather than paternal grandmothers. . . .

The Fischers' argument went beyond the drunken imprecations of old-timers in the bars of Goroka, but it depended on what had not yet been demonstrated in 1961: the presence of an infectious agent.

Mike Alpers remembers having lunch with the Glasses in Melbourne before he left for the NIH in early 1964. "We talked about all this, about the real possibility of cannibalism being involved." When he discussed the idea with Gajdusek, however, he found the American reluctant to concede the possibility. "It seemed to me that Carleton was embarrassed," Alpers says today. "He didn't want kuru associated with cannibalism because he thought the disease was already too exotic."

Gajdusek may well have been embarrassed by the idea of a connection between kuru and cannibalism, particularly since he had been welcomed by the Fore and their neighbors and had come to cherish them. But the more rigorous reason for Gajdusek's reluctance was scientific—in Alpers's words: "This practice in itself and by itself could not explain the epidemiological features of kuru; it required in addition the transmissible agent to do so."

With Elisabeth Beck's December 1965 report on the pathology of Georgette's brain—she found extensive loss of nerve cells, kuru plaques, astrogliosis and spongiform change so extensive she called it "severe status spongiosus" and noted that it "had completely disrupted the [structure of the areas it damaged] leaving very few nerve cells intact"—evidence that kuru was transmissible was now at hand. Gajdusek, Joe Gibbs and Alpers hastened to report their breakthrough, working up a paper for the British scientific journal *Nature* as soon as they received Beck's findings. By then a third chimpanzee was showing signs of cerebellar disease. The paper appeared in February 1966; it concluded with language intended to evoke Koch's postulates:

> This remarkable clinical correspondence of a disease developing successively in three chimpanzees each inoculated with brain material from a different kuru patient, the onset in each after a very similar long incubation period, the fact that there is no such syndrome of chimpanzees known to occur spontaneously or seen at present in our many control animals, and the remarkable similarity of the neuropathological findings, in the one case examined, to those observed in kuru victims lead us to believe that kuru has been transmitted experimentally to chimpanzees.

Robert Glasse explicitly connected kuru with cannibalism in a lecture to the New York Academy of Sciences in April 1967, but he also explicitly disavowed the authority of his evidence: "As a social anthropologist, I do not attempt to interpret these data in physiological terms, and thus have no specific hypothesis to offer relating cannibalism and kuru." In a paper published in *Lancet* the following year, Glasse, Shirley Lindenbaum and an Australian colleague finally made the connection explicit.

Nineteen sixty-seven was also the year the last Fore child under the age of ten died of kuru. For Alpers, that wonderful cessation of the grim death of children among people he had come to know and love, people who had begun naming their living children Michael in his honor, seems to have been a defining event:

> One morning, working at home in my makeshift office on the porch of our house in Bethesda, preparing a paper for a meeting of the International Academy of

> Pathology, I was waving all these possibilities around in my mind when suddenly the pieces of the puzzle clicked into place. Cannibalism, though it seemed to be a plausible explanation, did not on close analysis explain the phenomena as an independent mechanism. However, cannibalism as the single mode of transmission of the transmissible virus of kuru did make sense: it was suddenly all too painfully obvious. It was obvious because nothing further needed to be explained, and painfully so because of the agonies of uncertainty that existed when the explanations seemed so close at hand and yet not quite there: until that moment when it all clicked into place. But what a moment! Endocannibalism [cannibalism of relatives, that is] as the mode of transmission of the transmissible virus of kuru explained the sex and age distribution of the disease, since it was women and children who ate the infective brain and not the men; and the cohort of children growing up who were free of kuru had also grown up in a community now free of endocannibalism; since the practice was endocannibalism and only relatives were consumed the disease was familial in its distribution. Fortunately the disease was not transmitted vertically from infected mother to child; and that too was now clear.

By 1968, the connection between kuru and cannibalism was no longer in doubt. Alpers and others were even able to identify specific cannibal feasts such as the one I described in the first chapter of this book, which took place in the South Fore village of Ketabi in 1950, and to trace from there the fate of each participant. A tragic mystery had been solved.

But 1968 saw another discovery as well, one that enlarged the significance of kuru and cannibalism beyond a small group of people living in the New Guinea Eastern Highlands. In November 1966, Gibbs, Gajdusek and Alpers had inoculated a three-year-old male chimpanzee with homogenized brain tissue from a fifty-nine-year-old English male patient. The man had died, the scientists would report, after "an unremitting and progressive brain disease of eight-month duration" characterized by spongiform damage so severe that upon autopsy the brain appeared "collapsed to half of its former width." The chimpanzee developed a similarly progressive brain disease thirteen months later and the course of the disease was strikingly different from the standard course of kuru in chimpanzees. This chimpanzee was somnolent, blinded in half its visual field and weakened on one side of its body. Killed at sixteen months post-inoculation, it showed brain damage similar to kuru in the cerebellum but also severe spongiform damage to the higher brain areas of the cerebral cortex. With Elisabeth Beck and the British physician who had supplied the brain tissue, Gibbs, Gajdusek and Alpers reported in the American journal *Science* in July 1968 that they had now also successfully transmitted *Creutzfeldt-Jakob disease* to a chimpanzee.

CJD was not confined to New Guinea. It occurred throughout the world, it incubated for months or years before it revealed itself in symptoms, it was evidently one hundred percent fatal—and now it also had been shown to be transmissible. If it could be transmitted to chimpanzees, it could certainly be transmitted to other human beings. Surgeons and dentists might be at risk. So might their patients be, particularly patients receiving transplants of tissue from the nervous system.

Worse, so long as the CJD agent—the virus that caused the disease, if virus it was—remained unknown, it was impossible to say with certainty that the agent could only be passed through injection: kuru was evidently passed by eating contaminated tissue; scrapie was passed by eating afterbirth or picked up from land where infected sheep had grazed.

A fatal infection among Stone Age cannibals in distant

New Guinea was a tragic curiosity. But a transmissible fatal infection that turned up everywhere in the world with no obvious source for the infection was a risk to all humankind.

By 1968, the first phase of slow-virus studies was over: the human forms of these strange diseases had been shown to be transmissible. The next phase now began, driven by an urgency of concern for public health: a race to identify the disease agents so that they could be neutralized or at least contained. But they would not be contained. Even as determined research began to pry loose the secrets of the transmissible spongiform encephalopathies, ignorance and human error gave them terrible new opportunities to kill.

Sixth Connection . . .

SPECIES	DISEASE	PROGNOSIS
man	kuru (transmits to primates)	fatal
man	Creutzfeldt-Jakob **(transmits to primates)**	fatal
sheep	scrapie	fatal
mink	transmissible mink encephalopathy	fatal

Part Two

The Strangest Thing in All Biology

SEVEN
The Disease That Wouldn't Die

Edinburgh and London, 1953–1967

THROUGH THE EXCITING EARLY YEARS of kuru work, British researchers had plugged away at scrapie. As Bill Hadlow had realized when he saw Carleton Gajdusek's exhibit in London, scrapie closely matched kuru in how it spread and how it damaged the brain and it was a natural disease, accessible for study in animals. Dr. Alan Dickinson, a gangling, articulate zoologist and geneticist now retired from directing the Medical Research Council Neuropathogenesis Unit in Edinburgh, recalls with some disdain Gajdusek's whirlwind early tour of British research establishments. "Gajd came with his kuru films and a toothbrush and he may have arrived unannounced. I met him then and I saw a bit of these film studies in a dark room. And I remember thinking, really, this is a clinician going about it"—a medical diagnostic approach, that is, rather than basic biological research. "It tells you nothing. A clinician's approach to a problem like this has its value, but not a value of understanding the nature of these diseases at all." There was solid value in Gibbs, Alpers and Gajdusek's Patuxent work with chimpanzees, but basic research was just getting started in Gajdusek's lab.

The British effort in the 1950s and 1960s focused on isolating the scrapie agent. Looking for scrapie under a microscope worked no better than looking for kuru or Creutzfeldt-Jakob disease—the damage was visible, but not the agent—nor were efforts successful to grow the agent in culture. To smoke it out, British researchers attacked samples of scrapie-infected tissue with a battery of different chemicals and tests. Later work on the human transmissible encephalopathies built on this earlier effort. The record leaves no doubt that the British led the way.

How living beings reproduce was a mystery for thousands of years. The mystery only began to open up to understanding with the development of the microscope. One popular theory proposed that living beings passed on little copies of themselves. In the case of humans, the copy was called a homunculus—a little man. Some pioneer biologists studying sperm under crude early microscopes thought they saw preformed homunculi inside the sperm capsules, like miniature submariners riding to duty. The biologists supposed that the sperm docked with a female egg merely for the homunculus's nourishment: the little man settled in, ate egg yolk and grew. This "preformist" theory miniaturized the problem of how organisms reproduce, but didn't solve it—how was the homunculus assembled in the first place?

Hard work in science slowly established that what was passed on from parent to offspring was not a copy but information: a set of instructions that directed the assembly of a new bacterium, insect, fish, bird or human being. A great deal of information had to be transmitted to make anything so complex as a living creature. An encyclopedia of the human genome, for example, would comprise several thousand volumes. Cells are too small to be seen with the naked eye: how could so much information be stored in anything so small?

Instructions can take many forms besides spoken commands or words on a page. The hole in a birdhouse that keeps jays out but lets bluebirds nest is a kind of instruction. So are the notches in a key that line up the tumblers in a lock, or the two ends of a bar magnet that sort for magnetic polarity. Given the size constraints and the load of information, instructions for making new creatures evidently took the form of large molecules—the arrangement of their atoms presumably forming some kind of code. If cells are small, even large molecules are significantly smaller. A large molecule within a cell is comparable in size to a cat on a cruise ship. And at cat-and-cruise-ship scale, a grape could represent an atom.

During most of the first half of the twentieth century, the molecules that seemed to biologists to be the likeliest candidates for carrying genetic information were the proteins. There are hundreds of different proteins in the body, more than half its dry weight, each kind with its own unique three-dimensional structure. Complex assemblies of organic molecules, proteins do the body's mechanical and chemical work, serving as structural supports, as active agents such as hormones and in combinations such as muscle fibers. The capsule at the heart of cells that sheltered the genetic information—the nucleus—contained much more of a long, chain-like molecule called nucleic acid than it did protein. But nucleic acids, at least in the organisms biochemists studied in those days, seemed to be made up of the same proportions of various chemicals—and how could the same proportions of ingredients specify so many different creatures? Proteins, with their great variety as well as complexity, looked much more promising. So nucleic acid was dismissed as merely a structural support material and protein championed as the molecule of information. Experiments that showed that a substance could be extracted from dead and broken bacteria that would transmit a specific trait when mixed with living bacteria, and then showed that the substance was nucleic acid, were rejected or ignored.

James Watson and Francis Crick, in their 1953 report on

the structure of the nucleic acid DNA—the famous double helix—won so much attention (and shared a Nobel Prize with their colleague Maurice Wilkins) because they demonstrated that a nucleic acid was sufficiently complex to carry the genetic information. DNA's alphabet of four chemical bases hooked together in pairs as rungs of a spiral ladder gave a genetic code of sixty-four units (4 × 4 × 4)—sixty-one, as it turned out, to specify the twenty amino acids from which all the proteins of all living organisms are assembled, and three to signal the protein-assembling machinery when to stop. Nucleic acid has long since won the argument, but champions of protein genetics didn't give up their positions overnight. They fought the Watson-Crick model vigorously until experiment after experiment confirmed it.

The DNA double helix could make copies of itself (when a cell divided) by untwisting and separating the two matching chain-like sides of its "ladder." As the two sugar chains untwisted and separated, each took with it one half of the pairs of bases, attached to it like protruding broken rungs. Each base-studded segment of the chain, which spelled out the organism's genetic information in chemical code, then attracted matching molecules floating in the cell fluid; a new complementary chain assembled and the now doubled half-old, half-new ladders twisted into double helixes—making two DNA molecules where only one had been before, both faithful copies of the original, one to go with the daughter cell, one to stay with the parent.

But DNA alone turned out to be only indirectly responsible for protein assembly. Cells made proteins almost continuously, and another nucleic acid, RNA, single-stranded instead of double and using a different sugar-phosphate to make its chain, served as a crucial intermediary in the process. Cells didn't need the entire DNA sequence to make a protein. Each protein could be built from only a small part of the DNA sequence, a part known as a gene. A friend of

Carleton Gajdusek's, a distinguished Hungarian microbiologist named George Klein, who directs the Microbiology and Tumor Biology Center of the Karolinska Institute in Stockholm, explains the process in an essay:

> The nucleus of every cell in every organism contains all the information that each cell will need during the organism's entire life. But only the cornea of the eye is transparent, only cells of the retina have an apparatus that can respond to light, only kidney cells can filter urine from the blood, only muscle cells can contract to bend an arm or pump blood to the heart, only the red blood corpuscles can transport oxygen. . . .
>
> The functional diversification of cells is due to the fact that only about 2 percent or even fewer of all the genes are functioning in each cell. Different "specialized cells" use different parts of the genetic repertoire. . . .
>
> Every protein has a corresponding DNA instruction, and every set of instructions has a beginning and an end. A specially adapted protein—an enzyme—reads the instruction. But it has access to only a very limited part of the DNA, areas that are like small open windows on the facade of a big building. Different types of cells keep different sets of their windows open. The time and place for the opening and closing of all the windows are regulated during embryonic life by a meticulously precise program. The enzyme responsible for reading the instruction can also write while it reads. It makes a small strip of tape, a kind of telex message or quote. It is written in the language of RNA, which differs from the original DNA language only in some small details—a sort of dialect, so to speak. . . .

The master DNA molecule did not have to untwist completely to make strands of RNA. Instead, helper proteins pried open a short stretch of the double helix to separate the two chains, collected raw materials from the nuclear fluid and moved the opening along the DNA helix reading the code of bases on one DNA chain to assemble and extrude a strand of RNA. Behind the moving operation, the DNA chains twisted back together, safe from harm. When the RNA strand was complete, it was cut free. In a separate operation, a protein-assembly machine known as a ribosome then read its way along the completed strand of RNA, using the RNA as a template to run off copies of proteins.

Like the nucleic acids from which they were copied, proteins emerged from this translation process as chains. Such a protein chain might contain thousands of molecules linked together. After translation, a protein went through a second step to achieve its final form: it self-assembled. The various molecules of which a protein was made attracted or repelled different segments of the original long chain, folding it up into three-dimensional form like a rope tying itself into a complex knot. Most proteins had a structural component and an active component. The active component was usually globular, with grooves, holes, notches and bulges in its surface that fitted it tightly to other proteins or allowed molecules to dock with it to enhance or block chemical reactions. The structural component usually took the form of corrugated sheets. A finished protein might become the elastic of a muscle fiber or a pump for passing materials through a cell wall. It might become the oxygen-carrying component of a blood cell or a messenger moving information through a nerve. It might become an antibody that locked onto invading disease organisms and made them vulnerable to attack. Whatever its final form, a protein's shape was its destiny, determining the purpose it would serve.

After the long search for the molecule of genetic information, when the dust settled and the champions of nucleic acid over protein won the day, Francis Crick defined the central dogma of the new science of molecular biology: *DNA makes RNA; RNA makes protein.* Crick emphasized that the process was a one-way street so far as protein is concerned. There were "three transfers," he wrote in 1970, "which the central dogma postulates never occur": protein never made protein, protein never made RNA, protein never made DNA. But Crick was too careful a scientist to be dogmatic about his dogma. "Our knowledge of molecular biology," he added, ". . . is still far too incomplete to allow us to assert dogmatically that [this restriction] is correct." If its correctness had not yet been proven, it was nevertheless fundamental. Crick cautioned that the discovery of "just one type of present-day cell which could carry out any of the three unknown transfers would shake the whole intellectual basis of molecular biology. . . ." Out of all biology, Crick then drew from scrapie research to emphasize his point: "There is, for example, the problem of the chemical nature of the agent of the disease scrapie. . . ." The problem of the chemical nature of the scrapie agent was one the British work was exploring.

Since culturing scrapie-infected tissue had grown no bacteria and none was visible under the light microscope, everyone assumed the scrapie agent must be a virus. Bacteria are large, complex single-celled organisms capable of eating, excreting and reproducing themselves. Viruses are smaller and simpler. They were first distinguished from bacteria by their ability to pass through fine porcelain filters that screened bacteria out. With the development of the electron microscope it became possible to visualize them. They proved to be odd organisms, not much more than a strand of DNA or RNA coated with protein. A virus, the distinguished biologist Peter Medawar once quipped, is "a piece of bad news wrapped in protein." George Klein disagrees. "Viruses don't have to be bad news," he wrote me when I mentioned Medawar's quip. "Only a very minor portion of viruses are mad dogs. It's usually of no advantage to a virus to cause disease, much less to kill the host." One much-studied virus, a parasite of intestinal bacteria known as T4 phage, is essentially a tiny hypodermic syringe made of protein filled with DNA. When a T4 phage encounters its bacterial host, it docks and injects its entire DNA into the cell, after which the empty protein syringe breaks off and floats away. Viruses neither eat nor excrete and they can only make copies of themselves by hijacking the protein-assembly machinery inside cells.

If the scrapie agent was a virus, it was different from every other known virus. Sheep, like people, have elaborate immune systems for recognizing and destroying foreign protein such as the protein coats of viruses—but like kuru, scrapie caused no inflammation, raised no fever, evoked no immune response at all. Between 1939 and 1953, an Edinburgh researcher named D. R. Wilson had subjected the scrapie agent to many different tests, laboriously proving the results by injecting the challenged samples into sheep and waiting out the long incubation period to see if the sheep eventually sickened with scrapie. "He had not found a virus," writes one of his colleagues, "but he had established that the agent was highly resistant" to many kinds of insult. It survived thirty minutes of boiling. It survived two months of freezing. It survived disinfection with strong formaldehyde, carbolic acid and chloroform. It passed through fine filters and was small enough to stay in suspension even when spun in solution in a centrifuge at four hundred thousand rpm. It "remained viable in dried brain for at least two years, and resisted a considerable dose of ultraviolet light. It could be transmitted sheep to sheep" by injecting it into or under the skin, into a vein or directly into the brain. Diseases are usually consistent in the damage they cause, but scrapie produced fewer holes in the brains of experimentally infected sheep than in those of sheep that contracted the disease naturally. "In summary," writes the colleague, "by 1953 Wilson had produced evidence that this was a very eccentric transmissible agent."

Scrapie was so eccentric, unfortunately, that Wilson hesitated to publish his results, which were never widely known. Other British researchers took up the chase, sometimes repeating Wilson's work. William Gordon at Compton grandly assembled more than a thousand sheep representing twenty-four different breeds, injected half of each breed with scrapie brain and discovered over the next two years that different breeds had different susceptibilities to infection. Goats and sheep were successfully infected by feeding them scrapie-infected tissue, an oral route of infection that helped explain how the disease spread in flocks and supported the idea that cannibalism spread kuru. Goats and then mice demonstrated an inverse correlation between the size of the dose and the length of the incubation period—scrapie brain, with a higher concentration of the infectious agent, produced symptoms in a shorter time than scrapie muscle, with a lower concentration of the infectious agent. (Muscle from infected animals did, however, produce disease in other animals, an ominous discovery that everyone promptly forgot. Muscle, properly sliced and packaged, is what we call meat.)

Two conflicting British discoveries in the late 1960s intensified the controversy over the nature of the scrapie agent. One was the work of Dr. Tikvah Alper, a young woman conducting research in radiopathology at London's Hammersmith Hospital, and two of Gordon's colleagues at Compton, D. A. Haig and M. C. Clarke. The three researchers bombarded freeze-dried solutions of scrapie-infected mouse brain with high-energy electrons generated by a powerful linear accelerator. Electron bombardment disrupts molecules. Sufficient radiation can damage or kill an organism. Tikvah Alper and her colleagues knew from earlier research that there was a relationship between the intensity of the electron bombardment and the size of the target molecule: the more intense the bombardment, the smaller the molecule it would disrupt. That relationship gave them a way to estimate the size of the scrapie agent, whatever it was. After they had bombarded their samples, they reconstituted them, made a range of dilutions and injected the increasingly diluted samples into scrapie-free mice. The most dilute sample that still produced disease in the mice gave them a marker of biological activity. They then compared that dose to a table of comparable doses of biological agents of known size, including three phage viruses, a DNA molecule and three proteins. There results were striking: if the scrapie agent contained a nucleic-acid genome, that genome had to be a thousand times smaller than the genome of the smallest known virus.

In the second stage of their experiment, Alper and her colleagues mixed together a solution of phage virus and a solution of scrapie brain and exposed the mixture to strong ultraviolet light. UV light is germicidal; it kills germs by damaging their DNA. But a dose that reduced the infectivity of the phage virus to only one percent reduced the infectivity of the scrapie agent not at all.

These straightforward experiments led Alper, Haig and Clarke to a stunning conclusion, which they stated in plain English in the paper they published in 1966:

> Since the scrapie agent multiplies in the host animal, it has been assumed that nucleic acid must be a part of its structure. However, the evidence that no inactivation results from exposure to a huge dose of ultraviolet light, of wavelength specifically absorbed by nucleic acids, suggests that the agent may be able to increase in quantity without itself containing nucleic acid. This possibility is supported by the data from electron irradiations, since these yield a target size which is implausibly small as a nucleic acid code.

An infectious agent that reproduced without nucleic acid would be unique in all biology. This extraordinary possibility was the reason Francis Crick would example scrapie as a challenge to his central dogma. It was also the reason a disease of sheep in Great Britain and a disease that affected only a small population of former cannibals in the wilds of New Guinea became notorious among biologists throughout the world.

Around the same time, Alan Dickinson and his colleagues in Edinburgh pursued a second line of experiments that implied conclusions just the opposite of the Alper experiments. Dickinson was interested in exploring the differences between scrapie agents collected from different sources. To do so he began a long-term program of inoculating large numbers of mice and following carefully how they changed as the disease developed. He soon found that scrapie brain from different sheep, goats or other mice developed different incubation periods in mice. Some samples produced symptoms at 171 or 148 or 155 days; other samples required 328 or 466 or 602 days to reveal infection. These various incubation periods were remarkably stable, deviating by less than two percent when scrapie agent from the same source was passaged from mouse to mouse. Dickinson interpreted this discovery to mean that within the scrapie species there were different strains of the agent—"strain" meaning a variant form like the variant forms of flu virus that plague humanity from year to year, some of them more lethal than others and each producing a characteristic set of symptoms. Strain variations are common in microbes; Edinburgh researchers eventually identified more than twenty different strains of scrapie.

Dickinson and his colleagues found other differences between scrapie strains besides incubation period. Different strains also produced different characteristic brain damage—differences in the distribution of spongiform damage (holes in nerve cells) and in the accumulation of amyloid plaques.

Just as incubation period was stable for a given scrapie strain, so was the pattern of brain damage, which meant it was possible to determine, by looking at brain cross sections, which strain killed the mouse.

If a disease agent in the brain tissue of a sheep, inoculated into a mouse, produces a pattern of incubation and brain damage that can be passaged consistently from mouse to mouse, that must mean information is being passed from animal to animal. The molecules that carry genetic information are the nucleic acids (about half the viruses have RNA genomes instead of DNA). Tikvah Alper's work implied that scrapie had no nucleic acid. Dickinson's work seemed to demonstrate, to the contrary, that nucleic acid *must* be present in scrapie—to carry the detailed information necessary to perpetuate stable strains.

Further work by Dickinson and his colleagues argued even more strongly for the presence of strains. The Edinburgh researchers inoculated a mouse with a pure scrapie strain, 22C, that had a long incubation period, waited nine weeks to give the 22C infection time to take hold and then inoculated the same mouse with another pure scrapie strain, 22A, that had a short incubation period. When they studied the mouse's brain after it sickened with scrapie they found damage characteristic of the 22A strain, the one that had been introduced later. Which meant, they wrote in *Nature* in 1972, that the two strains had competed with each other (probably for the sites in the brain where the scrapie agent reproduced) and the 22A strain had won the competition—a Darwinian struggle for survival that offered even more dramatic evidence that the scrapie agents differed from each other.

Was scrapie an organism with nucleic acid or not? If not, what on earth was it made of and how did it preserve and transfer information? One more development during the later 1960s added further complexity to the confusion: researchers demonstrated that challenging suspensions of scrapie brain with enzymes known to damage nucleic acids produced no reduction in infectivity (just as Tikvah Alper had shown with UV light). But challenging matching suspensions with enzymes known to damage *proteins* reduced infectivity dramatically—by more than ninety percent. Was it possible that the scrapie agent was an *infectious protein,* whatever that might be? If so, it looked as if the central dogma of molecular biology would sustain serious if not fatal injury and the battle between protein and nucleic-acid genetics would have to be rejoined.

Or would it? Scientists working on the spongiform encephalopathies scrambled to find an explanation that would save the central dogma, on the sensible grounds that every other organism in nature followed the rules, so why should one small class of disease agents be the lone exception? "Everyone was making glib guesses," Gajdusek recalls. Alan Dickinson pointed out that Tikvah Alper and her colleagues had jumped to a conclusion. Their results didn't prove that scrapie lacked nucleic acid. Its nucleic acid might be protected from UV light and electron bombardment inside a particularly sturdy coat of protein. A French researcher at the Curie Institute in Paris, Raymond Latarjet, pointed out that scrapie was only twice as resistant to radiation as polyoma virus, which causes tumors in young mice. Polyoma protected itself from radiation by repairing the damage to its DNA; perhaps scrapie repaired its DNA as well.

Not a research biologist or a physician but a mathematician at Bedford College, London, suggested a remarkable and prescient alternative in *Nature* in 1967. J. S. Griffith mulled over the Alper work and proposed purely on logical grounds that under certain circumstances proteins might be able to self-replicate. "There are at least three distinct kinds of way in which [protein self-replication] could occur," Griffith wrote. He believed there was "no reason to fear that the existence of a protein agent would cause the whole theoretical structure of molecular biology to come tumbling down." One of his three ways didn't fit scrapie's behavior. The other two were not ruled out. Griffith's description of them is mathematical and technical, but they come down to two basic mechanisms. Both involve *host* proteins, so the genetic information that coded for them would be supplied by the host (the animal that got the disease), not the infectious agent—which would solve the problems that Tikvah Alper's work had raised.

Griffith's First Way would make the scrapie agent a protein which switched on a damaging reaction that was normally repressed in the host. If the protein was one of the many that are common to a number of different animals and organisms, it would work even if passaged from sheep to goats or from sheep to mice. If the repressed reaction in the host wasn't repressed with one hundred percent efficiency, it might sometimes switch on spontaneously. Scrapie outbreaks occurred, for example, that had no obvious source, and CJD, a comparable human disease with worldwide, low-level distribution, was thought to occur spontaneously.

Griffith's Second Way would make the scrapie agent an aberrant form of a normal host protein that sometimes got made spontaneously. If such an aberrant form served as a template, it might induce the production of more aberrant copies which might then accumulate until they caused lethal damage.

If scrapie followed either of these two Ways, Griffith concluded, "then it is a protein or a set of proteins which the animal is genetically equipped to make, but which it either does not normally make or does not make in that form. It may be passed between animals but be actually a different protein in different species. Finally, in either case, there is the possibility of spontaneous appearance of the disease in previously healthy animals."

Both Griffith's Ways provided a mechanism that might account for scrapie strains as mutations in the gene that coded for the host protein. Speculation isn't proof, but the British mathematician's ideas took root.

Commuting between Bethesda and New Guinea during these years, Carleton Gajdusek directed a zoological garden of kuru, CJD and scrapie studies and observed and contributed to the irrevocable changes that followed the opening of the Eastern Highlands to the outside world. "Human finger necklaces are completely gone in this part of the Kukukuku," he recorded matter-of-factly in his journal one day. In the village where the Fore had built him a house, Agakamatasa, he noted that "letter writing has caught on extensively" among Fore children even though they had received "no more than a year of minimal schooling." The range of his activities stimulated him:

> Teaching modern anatomy terminology and current therapy, worrying about warfare and possible attack, treating yaws. . . , collecting war raid chants and sing-sing. . . , inspecting men's houses for dangerous armament which might spell trouble, distributing marbles [for marbles games] to our carrier line [of boys who hauled supplies for him when he went on patrol], and translating Faulkner's short stories into Fore and Kukukuku, largely through Pidgin for my line . . . these incongruous and almost anachronistic activities are no stranger than purchasing recently used war maces. . . , stone axes and war clubs for bars of soap, needles, knives, and boxes of matches . . . [,] purchasing a meter-high, two-meter-diameter mound of kaukau [sweet potatoes, that is], corn, sugar cane, various greens. . . , pandanus nuts, and even native [plant-derived potassium] salt bars for spoonfuls of purified [salt], matches,

> soap, and small knives for a carrier line, and then arguing with a father about the removal of ceremonial belly bands to try to clean and dress matted crusts of impetigo covering the trunk of his ailing son . . . these contrasts are all wonderful, exciting, intellectually stimulating and I do not look forward to coming out to "civilization" tomorrow even though it be from one of the most remote of all government outposts of New Guinea!

Kuru was declining dramatically. It ceased among four-to-nine-year-olds by 1968, among ten-to-fourteen-year-olds by 1972 and among fifteen-to-nineteen-year-olds by 1973. Deaths among adults had declined by more than two-thirds, to fewer than sixty per year in 1972. No child born since the Fore gave up cannibalism had developed the disease. The adaptive Fore had taken the advice of their Australian mentors and planted coffee; nurturing their small coffee plantations beside the women's gardens gave Fore men respectable and profitable work now that intergroup warfare had ceased under Australian pacification. (Gajdusek thinks the Australians deserve a Nobel Peace Prize. "They pacified a more complex area than the whole of Europe, North America and Asia," he told me, "and they did it without bloodshed, with ceremonial law. They called crowds together, put on red robes and periwigs and by God they dispensed justice.")

Shirley Lindenbaum went out to New Guinea to follow up her earlier anthropological studies and found a changed world:

> When I returned to the South Fore for several weeks in 1970, signs of a new way of life abounded. Local groups around Wanitabe had built their own feeder link to the main road between North and South Fore, to facilitate the entry of coffee buyers from the towns of Kainantu and Goroka. By 1970 also, several hamlets in the South

Fore had their own canteens, stocking canned food, small goods, and clothing. These were local versions of the larger, white-owned trade stores that earlier had introduced Western goods into the Highlands. The accelerating demand for cash was such that it now cost more to purchase a chicken in the South Fore than it did in New York. Just as startling, I was awakened the first morning of my return by a small party of children shouting "Good morning" as they hurried to the primary school higher up the mountainside.

Whatever his personal preferences, Gajdusek did not neglect his expanding Laboratory of Central Nervous System Studies at the NIH in Bethesda. Joe Gibbs ran the laboratory, supervising a staff of talented younger researchers ("I was Carleton's jockstrap," Gibbs cracks). They reported various levels of success under various experimental conditions in transmitting kuru, CJD, scrapie and transmissible mink encephalopathy to chimpanzees; gibbons; New World monkeys including spider, squirrel, capuchin, woolly and marmoset; Old World monkeys including rhesus, bush baby, stump-tailed, cynomolgus, mangabey and African green; sheep; goats; calves; mink; albino and black ferrets; domestic cats; raccoons; skunks; mice; rats; golden and Chinese hamsters; gerbils; voles, guinea pigs; and rabbits. This work established the remarkably wide host range of the transmissible spongiform encephalopathies, the TSEs. It was justified, they wrote in the American journal *Science,* because there were "no indicators of the virus . . . available other than infectivity, clinical disease and neuropathological lesions in an experimental animal." They successfully passaged kuru from chimpanzee to chimpanzee, satisfying another of Koch's postulates. Looking into the phenomenon of strains, they identified no fewer than 112 different strains of CJD (each from a different human victim) with different incubation periods and disease patterns. If Alan Dickinson had thought Gajdusek's field work among the Fore amounted to no more than clinical studies, by the early 1970s no one questioned the status of the American's laboratory as the world center of TSE work. And so it was to Carleton Gajdusek and his team that the medical community turned when patients began dying.

EIGHT
High-Tech Neocannibalism

New York, 1971–1974

DISEASES DOCTORS UNINTENTIONALLY CAUSE are called iatrogenic, Greek for "physician-born." The first known human-to-human transmission of spongiform encephalopathy outside the Fore was iatrogenic. Dr. Arthur DeVoe, an eye surgeon and chairman of the department of ophthalmology at the College of Physicians and Surgeons of Columbia University, in New York, examined a fifty-five-year-old woman in 1971 who complained of seeing halos around lights and whose vision was clouded when she woke in the morning but improved during the day. That sounded like damage to the cornea, the eye's transparent window, and when DeVoe examined the woman he found her corneas waterlogged and swollen, a condition known for the Viennese physician who first described it in 1910 as Fuchs's dystrophy. (Light refracting through the swollen corneas caused the halos; the woman's vision improved during the day because the excess fluid responsible for the swelling could evaporate from her opened eyes.) Less severe cases of Fuchs's dystrophy sometimes respond to medical treatment, but cases that have progressed to waterlogging and swelling require corneal transplant, one cornea at a time.

A donor became available, a middle-aged man with a two-month history of memory loss and involuntary tremors who died of pneumonia. Down in the hospital morgue, an ophthalmologist harvested one of the man's eyeballs, immersed it in sterile saline in a small jar and delivered it to surgery.

DeVoe's patient was waiting, drowsy from sedation, her face veiled from him with a sterile drape with a round opening for one eye. DeVoe took up a syringe of Novocain with a long, thin needle and inserted the needle into the socket of the woman's eye, guiding the needle around and under the eyeball. He injected Novocain and the optic nerve and ciliary nerves went numb and the eye went numb and blind, temporarily blind.

An assistant relieved DeVoe of the syringe and passed him a scalpel and the donor eyeball nested in gauze. DeVoe sliced across the eyeball as you would slice the cap off a soft-boiled egg and inverted the staring cap into a Teflon cup. His assistant passed him a stainless-steel, razor-edged button-cutter then; twisting the cylinder with a practiced hand, the surgeon punched out the donor cornea from the inside.

When he knew he had a good replacement, DeVoe centered the button-cutter on the cornea of his patient's anesthetized eye, twisted it again to cut out the woman's diseased cornea and lifted cutter and cornea away. The hole opened into the eye shimmered with liquid.

Holding the donor cornea like a contact lens, DeVoe lowered it over the hole in his patient's eye. It fit perfectly. Meticulously, across the next hour, DeVoe joined the edges of the cornea and the woman's eyeball together with stitches of fine nylon thread, burying the knots in the wound.

The eye healed. The woman could see again clearly through the dead man's cornea and the operation seemed a success. But the optic nerve connects the eye directly to the brain, providing a channel for infection, and the brain of the man who died of pneumonia, who had not been autopsied until after his cornea was harvested, showed the characteristic damage of Creutzfeldt-Jakob disease. A year and a half after her operation, the woman began feeling nauseated, had difficulty swallowing, came to drool and stumble and jerk, went spastic, went mute, gradually introverted into vegetable oblivion. Two years beyond her surgery, emaciated and bedsore, she mercifully died. On autopsy her brain looked like the brain of the man who had donated his cornea—like a sponge. If Arthur DeVoe had only known before the transplant operation. A sickness had oozed from the cornea he'd implanted and eaten holes in his patient's brain.

Carleton Gajdusek and Joe Gibbs had already begun exploring the risks of surgical contamination with CJD when DeVoe's case came to their attention. In 1973 they had examined the brain of a fifty-four-year-old neurosurgeon who had died of a malignant skin disease but with signs of progressive brain disease as well. Brain tissue from the neurosurgeon inoculated into a chimpanzee and a squirrel monkey produced TSE (transmissible spongiform encephalopathy) in both animals. The two scientists reported the case at an international congress of neurology, pointing out that the neurosurgeon's CJD might have been spontaneous but might also "reflect a professional hazard to surgeons and pathologists who handle human brain tissue."

To alert professionals to that hazard, Gajdusek and Gibbs prepared a technical note for the *Journal of Neurosurgery.* They reviewed CJD transmissibility. They advised that it was reasonable to assume that the CJD agent was at least as resistant as the scrapie agent to heat, formaldehyde and ultraviolet light. "In particular," they wrote, "one must assume the agent is not killed by boiling." They pointed out that physicians often misdiagnosed CJD as Alzheimer's disease, as the form of cerebral atrophy known as Pick's disease, or as many other conditions including brain tumors and strokes. They recommended sterilizing instruments used on such patients in an autoclave—a machine used in hospitals that kills even hardy microorganisms with hot steam under pressure—for at least thirty minutes, twice the standard autoclaving time. They recommended treating all organs removed from such patients as infectious, even those fixed in formaldehyde. They had found only one chemical, chlorine bleach, that reliably killed the scrapie agent and they recommended using it to decontaminate floors and other surfaces where tissue might have fallen.

By the time the technical note was ready for publication, in the spring of 1974, DeVoe and his colleagues had reported on their iatrogenic corneal transplant in the *New England Journal of Medicine.* Gajdusek and Gibbs mentioned the tragic case in an addendum to their note. They contacted DeVoe and asked him for tissue from the dead woman's brain. By then it had been stored at room temperature in a four percent formaldehyde solution for seven months. Gibbs homogenized it and injected it into a chimpanzee. A year later, the chimpanzee sickened with TSE, proving that the case was infectious.

In 1963, Carleton Gajdusek had begun adopting children from New Guinea and Micronesia. He wanted a family and he wanted to help the premodern tribal groups that had welcomed him into their lives by educating some of their children. The first child he brought home with him was a twelve-year-old Anga boy named Mbaginta'o who had worked with him for years assisting with kuru examinations and autopsies. An old friend of Gajdusek's, Dr. Gunther Stent, professor of neurobiology at the University of California in Berkeley, remembers the occasion well. "Mbaginta'o arrived in San Francisco clad in his only post–Stone Age togs: white shorts, polo shirt and sandals. To get him into some street clothes more suitable for his onward flight to Washington . . . we took him to Berkeley's finest department store." In Washington, Gajdusek enrolled Mbaginta'o in an exclusive private day school. Mbaginta'o was the first of thirty-eight children Gajdusek would impoverish himself to import into the U.S. across the next thirty years, seven or eight at a time, and sponsor through high school and college. When Gajdusek established a branch of his NIH laboratory in a Level Four containment facility the U.S. Army made available at Fort Dedrick, in Frederick, Maryland, northwest of Baltimore, a wealthy friend loaned the maverick pediatrician an elegant eighteenth-century manor house sited on a hilltop estate outside town. The house, Gunther Stent remembers, filled up with children and "came to look like an ethnographic museum, chockablock with hundreds of artifacts Carleton had brought home, along with the kids, from his Pacific bailiwick: pots, spears, shields, masks and canoes (hanging from the ceiling)."

Besides the house in Frederick, Gajdusek still maintained his family home in Yonkers. Touring museums, art galleries and other cultural sites was part of his grand scheme for educating his charges, and he pilgrimaged to Yonkers every autumn with his current crowd of adopted sons to visit Herman Melville's grave, which he'd discovered neglected in a Yonkers cemetery. It was in Yonkers, early on the morning of October 14, 1976, that Carleton Gajdusek, now fifty-three years old, learned he had been chosen to receive the Nobel Prize in Medicine that year jointly with Baruch Blumberg, an American biochemist who discovered among Australian aborigines an antigen crucial to the development of a vaccine against hepatitis B virus. Gajdusek's Yonkers house was asprawl with eight of his adopted sons sharing sleeping bags with their girlfriends when television news crews began pounding on the front door. "Major reaction," Gajdusek wrote in his journal later that morning: "worry! Can I manage to complete the work I am trying to do, serve the boys well, and retain humility and creativity? . . . Mainly, can I live for my own curiosity and not for the adulation of others, or when others are in mind may they be as yet unborn and yet may I live for their adulation in humility?"

It was satisfying, he wrote, "to have become a Nobel Laureate while living in the house in which I first dreamed of medical research in microbiology as my vocation. Paul de Kruif's heroes' names are on the attic stairs where they remain today. . . . To be sharing this unanticipated news with friends . . . and with eight of my boys is a very welcome circumstance. However, to have the press so intrusive upon our private lives is intolerable and extremely disconcerting. . . . The house is loaded with TV people and reporters who are difficult to clear out." His thoughts turned to his colleagues. "I am very uneasy about the dreadful effect of the honor I receive on Joe Gibbs and Vin Zigas who were surely the two I would most logically expect to share it with. I have feared that Hadlow and some of our British or Scottish scrapie colleagues might be my 'running mates' if I ever received the award and that, to the exclusion of Vin and Joe, would have been less tolerable. . . . The only solution is to further our goals and elucidate the nature of the scrapie agent and the natural history of transmissible virus dementias. These two matters solved would exceed any achievement we have made thus far. I am sure that a full unraveling of the structure of the unconventional viruses would warrant another Prize."

Gajdusek stashed his gang of kids with his mathematician friend Benoit Mandelbrot, the discoverer of fractals, at IBM Research in Yorktown Heights and shuttled to the NIH long enough to speak to his delighted colleagues. Joe Gibbs introduced him proudly; if Gibbs was disappointed he didn't let on. The letter Gajdusek received announcing his Nobel informed him that his share of the prize would amount to $79,742, money he would spend on his adopted sons' education. (The Nobel endowment had declined sharply in value in those years, worse luck for winners; it has since been restored.)

Two days later, the Gajdusek ménage completed its pilgrimage to Melville's grave; Gajdusek had trouble finding it in the crowded cemetery until he turned his sons loose with their tracking skills. He'd had time to think about the justice of the prize by then, musing in his journal on October 16: "I see now, as I honestly rethink these last two decades, that the impetus for establishing slow-virus disease as a mechanism in human disease pathogenesis was far more my own than . . . even Joe's, and it is this that the Nobel committee noted. Hadlow had a part in it, so did Elisabeth [Beck], Igor [Klatzo] . . . and Joe Smadel, so did [William] Gordon and [John] Stamp [the director of the Moredun Institute] and Pal Palson and Björn Sigurdsson [both of Iceland] and for many years the work rested on Vin and later on Joe. To share the award with them would have been kinder." He shared the award with Baruch Blumberg instead, the Nobel committee citing both men "for their discoveries concerning new mechanisms for the origin and dissemination of infectious diseases." Twenty years before their triumph, even before he encountered the Fore, Gajdusek had speculated in his journal on the Western fascination with premodern societies:

> Primitive man has attracted the imagination, curiosity and romanticism of men of Western culture for centuries—even millennia. . . . Civilized man may see in the surviving remnants of primitive cultures his own origins and realize intuitively in observing them that other solutions to the problem of society and communal life are possible than those which historical events have selected to be his own lot. . . .
>
> The South Pacific, particularly, has been the source of romantic yearnings, and a large number of gifted artists and scientists have voyaged there from Europe and America in search of Rousseau's Noble Savage. . . . Pierre Loti, Robert Louis Stevenson, Joseph Conrad,

> Paul Gauguin, our Americans, Henry Adams and Herman Melville, and, more recently, Somerset Maugham and James Michener, have told the tales of the South Pacific with brush and pen more dramatically, if less exactly, than Bronislaw Malinowski, Alfred Russell Wallace, Charles Darwin, and Captain Cook. . . .

Both Gajdusek's and Blumberg's discoveries were gifts annealed in the suffering of premodern peoples; in pairing the two scientists for that year's prize, the Nobel committee may have been acknowledging the Rousseauian benefaction.

In the midst of preparations to travel to Stockholm, on November 28, Gajdusek heard from a Swiss physician, Dr. Christopher Bernoulli, that two of Bernoulli's young patients in Zurich had developed what appeared to be Creutzfeldt-Jakob disease following diagnostic surgery for epilepsy. "The letter from Dr. . . . Bernoulli . . . is so surprising and upsetting that I can hardly believe it," Gajdusek wrote in his journal that day. The instruments of contamination were evidently silver electrodes that Bernoulli had inserted into his patients' brains and left indwelling for several days to triangulate the location of their seizures prior to corrective brain surgery. The Swiss physician traced the contamination to a sixty-nine-year-old woman he had treated for CJD in 1974. He had used the electrodes to monitor her brain waves and sterilized them with seventy percent alcohol and formaldehyde vapor. A few months later, a twenty-three-year-old woman presented herself for treatment of drug-resistant epilepsy. Bernoulli implanted nine electrodes in her brain in November 1974, two of which had been used previously on the CJD patient, and on the basis of their readings performed brain surgery that freed her of her seizures.

But the young woman returned to see Bernoulli again in July 1976. By then she was six months pregnant and was having difficulty walking, felt nauseated driving, had trouble speaking and was experiencing memory loss. Bernoulli found eye jerks and loss of coordination. "She gave imprecise answers," he wrote later, "and looked 'giddy.'" Her deterioration was rapid. In mid-September she delivered a normal child by caesarean section. Since then she had become comatose.

Bernoulli's second patient was a seventeen-year-old boy with a similar history. The boy was close to death when Bernoulli wrote Gajdusek. "The physicians are worried lest a disaster be upon them," the American noted in his journal, "and they solicit my help. I have cabled them tonight that we stand by to study brain and other tissues from all their cases. . . ." The seventeen-year-old died five days later; an autopsy of his brain confirmed CJD. Tissue from both patients, implanted into primates, eventually proved transmission.

To the Nobel ceremonies in December, Gajdusek took a gang of his kids including Mbaginta'o, now twenty-five years old and answering to the first name Ivan, and seven younger boys from New Guinea, Micronesia and Singapore, sending ahead exact measurements for their striped trousers and cutaways. Joe Gibbs, Paul Brown, Gajdusek's secretary Marion Poms and two other friends joined the group as well. Gajdusek informed his hosts that the boys wouldn't need separate hotel rooms: they were used to sleeping on the floor in sleeping bags and could share his suite. Short of stature but strong, handsome and polite, they made an unforgettable impression.

So did Gajdusek. It was a Nobel tradition that each laureate should deliver a forty-five-minute lecture on his prize work. Unable to contain himself, Carleton Gajdusek spoke for two hours. "It was all right," George Klein told me, "because what he had to say was fascinating." In the course of his illustrated lecture, Gajdusek held up a plastic tube that radiation had blackened and pointed out that the scrapie agent was still alive inside and infective. He noted that even in partially purified scrapie samples, no viral particles could be identified under the electron microscope. He noted that no infectious nucleic acid was demonstrable from these unconventional diseases, nor any foreign proteins. He concluded, thinking perhaps of the Fore gift of suffering, that the investigation of the history and epidemiology "of a rare, exotic disease restricted to a small population isolate—kuru in New Guinea—has brought us to worldwide considerations that have importance for all medicine and biology." There was much left to know, Gajdusek concluded candidly, much that was "still inexplicable."

No more cases developed from the Swiss disaster, but Gajdusek, Gibbs, Paul Brown and others in Gajdusek's lab collaborated during 1977 on a special article for the *New England Journal of Medicine* warning medical practitioners to take special precautions working with CJD patients. Although they thought isolation rooms were unwarranted, they increased the recommended autoclaving time for medical instruments used in CJD cases from thirty minutes to an hour and advised against using donations of blood or tissue from demented persons. The ominous possibility that an irreversible, one-hundred-percent-fatal brain disease might be spread by blood transfusion was confirmed in 1978 in guinea pigs infected with Creutzfeldt-Jakob disease by researchers at Yale University. Gajdusek's lab started testing other body secretions and excretions—tears, saliva, feces, urine, semen and vaginal mucus—for infectivity. In 1980, Gibbs, Gajdusek and several colleagues reported transmitting kuru, CJD and scrapie orally to squirrel monkeys fed infected brain, kidney and spleen. In 1982, the New York State Institute for Basic Research in Developmental Disabilities, on Staten Island, reported transmitting scrapie by applying brain homogenate to the gums of mice. Scratching the gums with scissors first, or pulling a tooth, produced one-hundred-percent transmission; if the gums were left unscraped, transmission was still seventy-one-percent—emphasizing the possible risk of CJD transmission from dental work. At about the same time, as if to underline the danger, two researchers at Oxford University reported finding three cases of probable iatrogenic transmission of CJD by neurosurgery dating back to the 1950s and a more recent "cluster of three cases in eastern England possibly connected by dental procedures"—a dentist and two of his patients who had all died of CJD.

Despite such suggestive incidents, when Paul Brown reviewed the evidence in 1980, he concluded that except for "a few proven or highly suspicious cases of iatrogenic surgical transmission of CJD," there had been no documented instances of transmission among humans though any conventional points of virus entry: not by airborne, oral, venereal, skin or skin-puncture infection. But he cautioned at the same time that his review left him feeling as if he were going through "a maze from which the exit remains terribly obscure." The maze would soon prove crueler.

NINE
Infecting the Children

Edinburgh, 1976 / Palo Alto and Bethesda, 1985

ONE NIGHT IN OCTOBER 1976, at about the same time Carleton Gajdusek learned he had won the Nobel Prize in medicine, Edinburgh geneticist Alan Dickinson was shaken by a terrible premonition. "I had gone to bed," he told me, "and I think I must have been musing. I usually fall straight to sleep and then wake up sometimes a few hours later. After a short sleep, you're fresh and you can design experiments. But that evening I sat up in bed and thought, *Good grief, I wonder if they realize that the way they're extracting growth hormone*—for injecting into thousands of children throughout the world to treat dwarfism—*can concentrate the CJD agent.*"

Human growth hormone was first isolated in the 1950s from the pituitary gland, the small master gland attached to the base of the brain that secretes hormones that regulate growth, metabolism and reproduction. When a research physician demonstrated in 1958 that growth-hormone injections could stimulate a child with pituitary dwarfism to grow to normal height, pediatricians clamored for a supply of the hormone. The only source at the time was pituitary glands removed from cadavers in the course of autopsies. In the U.S., a National Hormone and Pituitary Program was set up within the National Institutes of Health in 1963 to collect cadaver pituitaries, extract growth hormone and distribute it to pediatricians willing to participate in experimental programs of treatment. Similar programs began at about the same time in other countries, notably Britain, France and Australia. "The NIH hormone was not made for profit," Gajdusek comments, but across the next twenty years "about fifty thousand children in the U.S., as many Europeans, as many Asians, and South Americans—perhaps almost two hundred thousand throughout the world—received growth hormone."

When Alan Dickinson realized that the procedure the British Medical Research Council (MRC) was using to concentrate growth hormone from cadaver pituitaries might also concentrate the CJD agent, he alerted the council to the potential risk. The MRC responded by commissioning Dickinson's institute to test the procedure—deliberately to contaminate a batch with CJD, extract growth hormone and inoculate it into test animals. The testing process was necessarily slow, since the incubation period for spongiform encephalopathy is long even in mice; results would not be available until 1983.

By 1976, according to Dickinson, genetically engineered growth hormone was on the horizon, not yet perfected but believed to be two or three years down the road. Its purity was guaranteed: it was produced by recombinant yeast with the genetic code for human growth hormone spliced into its DNA. In the meantime, at great expense, the MRC had accumulated large quantities of cadaver-derived hormone to meet British needs until the engineered hormone became available. The MRC weighed the theoretical risk of using the stockpile hormone for two or three more years against the real risk of stopping treatment of children afflicted with dwarfism and decided the gamble was worth it. "The treatment was immensely satisfying, you see," Dickinson says bitterly, "because you're not dealing with a condition which is life-threatening, you're dealing with something which enormously modifies the quality of life. Immensely satisfying, and against that you set this nebulous inconceivable entity, Creutzfeldt-Jakob disease, and the name is even difficult to pronounce, and who are these people at this animal research institute raising a fuss? What a joke. Please." The MRC decided to continue distributing cadaver-derived growth hormone while Dickinson and his colleagues checked out the purification protocol. Other countries were not even alerted to the risk.

The Edinburgh scientists eventually reported detecting no CJD in the hormone they purified. Paul Brown has disputed this conclusion. Following standard practice in their experiment, Dickinson and his colleagues inoculated their test animals with only part of the growth hormone they made. Brown argues that finding "an *undetectable level* of virus did not necessarily mean *no* virus, as subsequent events were to show. To demonstrate the total absence of infectivity, it would have been necessary to inoculate the total volume of material. . . ." Brown's point is that the CJD agent may have been lurking in the portion of the test batch that Dickinson's group discarded. The Edinburgh scientist defends the quality of his institute's work. "We just used common sense. In our scrapie research we had to keep strains of scrapie apart, and if you can do that, you can do anything." Dickinson's group may have been skilled enough to purify contaminated pituitary extract using the MRC protocol. The technicians at other institutions responsible for the MRC's growth-hormone stockpile evidently were not.

One of the first U.S. children to receive growth-hormone therapy was a California boy known to medical history as JRo. JRo suffered from multiple hormone deficiencies: not only growth hormone but also thyroid and insulin. He started on growth hormone in 1965, when he was two years old. His insulin-dependent diabetes made it difficult to treat his growth-hormone deficiency in the customary way, with thrice-weekly injections. "Thus," writes his physician, Stanford University pediatric endocrinologist Dr. Raymond L. Hintz, "JRo became a pioneer in the receipt of daily GH [growth hormone, that is] in fairly large doses over a long period of time." By his twentieth birthday, the boy had grown to a height of about five feet four inches. His family and his doctors considered him a medical success.

Paul Brown takes up JRo's story: "According to his mother, he was always a happy, optimistic and talkative youngster; perhaps because of his tenuous medical condition, he was also somewhat of a stoic. Therefore, when he complained, it behooved people to listen." In his twentieth year, JRo complained of dizziness on a flight with his family from San Francisco to Maine to visit his grandparents. His mother thought his blood sugar might be low, a common problem for diabetics. She gave him some candy and he seemed to feel better. But in Maine, Brown writes, "he turned down an offer to go for a spin on the lake in his grandfather's motorboat, saying that 'he didn't need to go for a spin because he was already dizzy.'"

JRo's dizziness persisted on his return to San Francisco. "When he got off the plane," Hintz reports, "his mother noted that he was clumsy and had awkward movements." She called Hintz on Sunday, June 17, 1984, and he arranged to examine JRo in the university hospital emergency room. He found the young man's speech and coordination impaired and his eyes twitching and arranged for him to see a neurologist. Hintz describes the brutal subsequent course of JRo's illness: "A series of disastrous neurological events . . . progressed rapidly over a few weeks' span. The parents were obviously alarmed at his downward course. . . . He developed severe dementia. . . , deteriorated to the point that he needed hospitalization, and died within six months of onset of symptoms. . . . On autopsy, the classic pathological [signs] of spongiform encephalopathy diagnostic of CJD were found."

When Hintz learned of the autopsy results, in February 1985, he was greatly disturbed. He remembered a conference he'd attended three years previously at the Food and Drug Administration (FDA) in Washington where slow-virus diseases had been discussed. "Because CJD was such an unusual disease in a young person," he writes, "I was very concerned that the CJD of JRo might have come about from contamination of pituitary GH supplies by the CJD infectious agent. If this were true, all of the patients that we had treated with pituitary GH . . . might be at risk of developing this disastrous degenerative neurological disease." In late February, Hintz drafted a letter about JRo's fatal illness and sent copies to the FDA, the National Hormone and Pituitary Program and the NIH. He proposed in the letter "a careful follow-up of all patients treated with pituitary growth hormone in the past 25 years. . . ."

The NIH administrator responsible for the growth-hormone program was Dr. Mortimer Lipsett. When Lipsett read Hintz's letter, he wasted no time. "I've never seen the government respond to something as quickly as they responded to Hintz's letter," Brown told me. "It was remarkable. Lipsett really leaped on it." The NIH administrator called a meeting of officials and scientists on March 8. They recommended alerting physicians to the possibility of a problem and canceling all nontherapeutic use of the hormone. They did not recommend suspending hormone treatment until JRo's CJD diagnosis could be verified and other possible sources of his exposure to CJD excluded.

"I remember the next six weeks as a stressful time," Hintz writes. "Although my colleagues at Stanford were fully supportive of my action, many of my other pediatric endocrine colleagues clearly felt that I was an alarmist, and they did not hesitate to tell me so. The pressure increased when it appeared that the therapeutic use of GH in the United States might be halted because of this apparently unique case. Thus, more than 3,000 patients with GH deficiency would no longer be able to receive needed treatment because of my single case report."

Lipsett scheduled a second, larger meeting at the NIH for April 19 to discuss whether or not to halt therapeutic use of growth hormone. He invited administrators, scientists, physicians and suppliers from several countries to attend. Hintz arrived in Bethesda on the afternoon of April 18 and learned that two more probable cases of CJD had been identified. "By an incredible coincidence," he writes, "these pituitary GH–treated cases had died of neurological disease within the previous few months."

Neither case had been reported earlier because a diagnosis of Creutzfeldt-Jakob disease had not been considered: both patients were too young to fit the usual CJD profile. One was a Dallas music-store clerk, thirty-two years old, whose onset of slight staggering his mother had been the first to notice. His family doctor had suspected inner-ear problems or a brain lesion but had found neither. The man had continued working as his loss of coordination progressed until a physician customer had suggested he see a neurologist. When he did, he was diagnosed with probable multiple sclerosis. Unlike JRo, he had shown no signs of dementia; Brown writes that the man "was still playing the game of 'Trivial Pursuit' until shortly before his death in February 1985."

The second patient, a Buffalo, New York, man, twenty-two years old, found himself veering to the right while out walking one day. Balance problems and double vision sent him to a physician, who told him he was malingering. He got worse, Brown reports, was diagnosed at Buffalo General Hospital with severe cerebellar syndrome and was discharged to a nursing home. Like the Dallas patient, he had minimal dementia up to the time he died—a pattern more like kuru than classical CJD.

When they heard of Hintz's case, the Stanford pediatrician recalls, the two doctors who had treated these patients reviewed their records "and realized that they were strikingly similar in their clinical presentation and inexorable fatal course. . . ."

"That was the death of pituitary-derived growth hormone," Brown says. At the NIH meeting on April 19, some of the foreign participants who supplied hormone to the U.S. program argued that U.S. purification methods must be faulty and that the problem was therefore strictly American. Taking no chances, Lipsett ordered all growth hormone treatment in the U.S. halted immediately, whatever the source of the hormone.

"The pediatricians were very unhappy," Brown observes. "They held out for as long as it was possible to hold out. They had a point when there was just one case. Two additional cases and that was that. But even then, the pediatricians were unhappy." Fortunately for their patients, recombinant growth hormone became available within a matter of months. "I don't know what would have happened had the recombinant hormone not been ready when it was," Brown adds. "My guess is that we would have continued native growth hormone for life-threatening situations, such as serious diabetes, and not given it to anybody else."

After the dust settled, Brown reviewed the risk that cadaver pituitaries might carry CJD. Since autopsies that included the brain were more likely to be performed on people who died of neurological disease, he estimated that one in a thousand pituitaries collected for the growth-hormone program probably carried the CJD agent. "Pituitary pools that contained 10,000 glands, therefore, had a good chance of containing from one to ten infected glands. With the further knowledge that approximately one-half million glands were processed between 1966 and 1977, it follows that between 25 and 250 infected glands would have been included in pituitary batches collected during this period alone."

The British reviewed their records when they learned of the three U.S. deaths and identified CJD as the cause of death of a twenty-two-year-old English growth-hormone recipient in February 1985, the same month as the Dallas music-store clerk. Now deeply worried, Brown, Gajdusek and Gibbs warned the medical community in a September article in the *New England Journal of Medicine* of a "potential epidemic of Creutzfeldt-Jakob disease from human growth hormone therapy." They tried to estimate the degree of risk, but could offer at best a range—from the four patients already identified to the worst-case epidemic of their warning. Given the long TSE incubation period, it was still too soon to tell.

For almost a year, no more CJD appeared among growth-hormone recipients. Then, in the spring of 1986, twelve years after her last treatment with growth hormone from the U.S. program, a young New Zealand woman died of CJD. A physician at Cornell Medical Center in New York City looked back over the records of the hundred patients she had treated with growth hormone and discovered one who had died of pneumonia in 1979 whose brain showed CJD pathology. A Pennsylvania patient died in 1987. More British cases began turning up in 1988. French patients began dying in 1989. By 1996, Brown tallies, the death toll stood at about eighty cases worldwide: "Fifteen in this country, two in New Zealand and one in Brazil who got hormone prepared in this country. One in Australia [from] local hormone. Seventeen in the United Kingdom and about forty-odd in France."

Eighty deaths of previously healthy young people was bad enough. "It was only luck," Carleton Gajdusek grimly told a bioethics conference in Japan in 1992, "that saved thousands of people when we contaminated, inadvertently, the human growth hormone, prepared from human cadaveric pituitaries and injected into our children in an instance of high-tech neo-cannibalism. . . . Only fortuitously was it not more like two hundred thousand iatrogenic deaths!"

In their September 1985 article in the *New England Journal of Medicine,* Brown, Gajdusek and Gibbs had confronted squarely the larger danger of iatrogenic CJD contamination:

> We are once again dramatically reminded that human tissues are a source of infectious disease, and that any therapeutic transfer of tissue from one person to another carries an unavoidable risk of transferring the infection. In this context, we must continue to worry about such products as follicle-stimulating hormone, luteinizing hormone, prolactin, and human interferon, as well as skin, bone, bone marrow, dura mater, blood vessel and nerve grafts and organ transplantation.

True to prediction, the first CJD death from transplanted dura mater—the tough outer covering of the brain, used for patching after brain surgery—showed up in 1987. Six further cases were reported from New Zealand, Britain, Italy and Japan in the next five years. In five of the seven cases, the dura mater traced back to one German firm, which processed the tissue with hydrogen peroxide and radiation. During the same period, a hormone used in infertility treatment, gonadotrophin, infected with CJD and killed three Australian women. A fifty-one-year-old Dutchman died of CJD two years after heart membrane was used to repair his perforated eardrum, but whether the CJD was iatrogenic isn't certain since the donor wasn't autopsied. "Iatrogenic CJD," Paul Brown concluded in a 1992 review, ". . . is a serious and growing problem in the high-technology milieu of modern medicine."

Brown has a sardonic name for these continuing human tragedies. He calls them "friendly fire."

TEN
A Candidate for a Modern Wonder

Staten Island, Agakamatasa and San Francisco, 1972–1985

WHILE MEDICAL DISASTERS ACCUMULATED in the 1970s and 1980s, research into the nature of the scrapie agent advanced. Researchers hoped that understanding the transmissible spongiform encephalopathies might open the way to understanding other obscure diseases of the central nervous system that killed far more people, notably multiple sclerosis and Alzheimer's. They worried as well that not to know the cause of a disease was to risk its full wrath if it ever broke loose. It isn't necessary to identify a disease agent to prevent a disease—many diseases have been controlled with improved sanitation or medication before their causes were found—but the TSEs were exceptionally elusive.

Gajdusek and Gibbs extended their operations from the NIH and Fort Dedrick to an outsource animal facility in Louisiana. "They were even inoculating this stuff into alligators!" one skeptical colleague recalls. They funded work at the Rocky Mountain Laboratory as well. The freezers filled with frozen brains and the papers poured forth, dozens every year with the Nobel laureate among the authors, adding bit by bit to the complex picture. But the picture was still out of focus, the subject still confused. When Gajdusek had speculated in his journal on the justification for his Nobel, he'd predicted that "a full unraveling of the structure of the unconventional viruses would warrant another Prize." He was not alone in that opinion. Elsewhere in the U.S., in Britain, Europe and Asia, other committed scientists—some almost monomaniacally ambitious—took up the search.

Patricia Merz, an intense young woman who had not yet found time to finish her Ph.D., scored a major breakthrough in the late 1970s wielding an electron microscope at the New York State Institute for Basic Research in Developmental Disabilities on Staten Island. Five feet tall, a fanatic skier, solid and compact, with strong hands, blue eyes and thick brown hair piled on her head like a brioche, Merz got into TSE research following her husband-to-be to Staten Island from graduate study in Philadelphia. "The Institute has just been formed in 1969," she told me in her Staten Island laboratory. "It was set up to look at brain diseases and particularly those that infect children. The TSEs were just coming on for research at that time, the next round of maybe giving a hint of what's going on with the nervous system."

Since the various TSEs behave in similar ways and cause similar damage, studying one might reveal basic information about them all. Merz got interested in scrapie, the TSE of choice for laboratory research because it had been studied most extensively in stabilized lines of laboratory mice and hamsters. No one had yet seen the scrapie agent. Everyone was still tracing it by laboriously following its infection in laboratory animals. A jigsaw-puzzle aficionado, pragmatic and grounded and impatient with abstraction, Merz decided to go looking for the agent itself. To do that, she first had to teach herself to operate the electron microscope and to identify the tissue structures that superb tool revealed.

A machine the size of a rolltop desk, the electron microscope uses a beam of electrons rather than light to illuminate a specimen and magnifies by focusing its beam with magnets rather than glass lenses. Because electrons have much shorter wavelengths than visible light, the EM can achieve much higher magnification than the light microscope—up to hundreds of thousands of times the specimen's original size. That means the EM can visualize much smaller specimens, as small as small viruses and broken fragments of protein or DNA. EM visualization can be direct to a fluorescent screen like a television screen (but usually glowing only one color, green or orange), or photographic. Since the screen is relatively dim, the EM has to be isolated in a dark room. "It's sensory deprivation," Merz explains. "You see things that aren't there—that's why you take pictures."

Besides teaching herself to operate the EM, Merz had to learn staining. Tissue staining for the EM is different from tissue staining for the light microscope. Rather than *coloring* specimens selectively, EM stains *thicken* specimens selectively, coating them with compounds of heavy metals such as uranium that make them more opaque to electrons, so that they cast darker shadows on screen or film. Merz trained herself to sort out what such staining revealed. She would be homogenizing brain tissue, diluting it, coating a carbon grid with it, staining the dried residue and studying the highlighted debris of broken cells, protein, DNA and unidentifiable junk. The grid she used was the size of a small letter "o," divided up like a window screen into four hundred or more segments, and she could look at no more than one segment of the screen at a time. Yet that minuscule peephole would reveal a world of subcellular objects—the scattered raw materials, the machinery, the broken walls, the forklifts and pumps and finished products of the miniature factory that is a cell. "You get to know what a cell membrane looks like, what a nucleus looks like," Merz told me. "You can study photographs in books, you can study this and study that. But if you deal with fluids and particulate material and you stain them, you have absolutely no guidelines, you have to spend a long time building up frameworks of what things are, what they could be, what they might be, is it real, is it debris, is it whatever it is. Just you and the microscope. There's nothing else that guides you."

Merz was a natural at EM work, but teaching herself that work took two long years. By 1978, she was ready. She was collaborating by then with an English colleague, Robert Somerville. Other researchers had established that the scrapie agent was difficult to separate from the membrane—the outer wall—of nerve cells and might even be a component of cell membrane itself. Somerville was studying one such component. Merz offered to look at Somerville's samples under the EM. It would be a fishing expedition. She didn't know what she'd find. "Maybe I'd see something that was different. I was using a Hitachi 8, a very small EM." Into the darkened room with transformer hum and a glowing green screen for company. "The very first time I went looking, in February 1978, I shot a bunch of pictures. I developed the prints and lo and behold there were *sticks* on the prints, which I had not seen on the screen. Tells you how blind you can be. A very good example of how blind you can be. We didn't know what they were. They were only in scrapie samples. Robert didn't believe they meant anything."

The "sticks" lying scattered among the cell debris in Merz's photographs looked like short, broken lengths of twisted thread. Some had two filaments twisted together; some had four. They were almost inconceivably small: to enlarge them to an inch in length in photographs, Merz had use a more powerful EM and turn up its magnification to ×68,400.

She was living in the EM room now. She continued to notice that the sticks showed up only in scrapie samples. "They verified across all strains. That was the first time anything like that had happened. They were in every strain. They increased in quantity in the course of the disease. I was looking at coded samples, so I didn't know whether the sample was infected or not, but I could pick out the infected samples because of these sticks and I could tell you which scrapie strain it was just by the way they looked."

Like all good scientists, Merz attended scientific meetings to share her work and keep up with her field. She began showing her stick photographs around, asking colleagues what they thought the sticks might be. In particular, she wondered if they looked like the protein fibers that clump together to make plaques in the class of diseases known as amyloidoses. (Igor Klatzo had found amyloid plaques in the brains of the kuru victims Gajdusek had sent him in the late 1950s. Amyloid plaques are diagnostic of Alzheimer's disease. Amyloid is also produced in the body in response to major infection; when obituaries report someone dying of "complications," the complications are frequently deadly deposits of junk amyloid, which can kill by clogging vital organs such as the liver or the kidneys. Robert Koch, of Koch's postulates, had discovered amyloid while studying tissue samples under the light microscope; he named it from the Latin word for starch, *amylum,* because the fine grains of amyloid reminded him of the gritty starch grains that stipple the cores of pears. The "grains" Koch saw in his tissue cross sections were eventually identified as the cross-cut ends of amyloid filaments.) But none of the experts Merz consulted thought her sticks were amyloid.

Whatever they were, they were abnormal: they never turned up in healthy brain and they always turned up in scrapie-infected brain. Merz and Somerville eventually reported their discovery in a German pathology journal and named the sticks "scrapie-associated fibrils": SAF. In that first report, Merz stubbornly insisted on a connection between SAF and amyloid even though the experts disagreed. The two entities looked similar, she wrote. The acid test of amyloid is its green glow under polarized light after staining with Congo red. SAF had failed that test, Merz reported candidly, but she

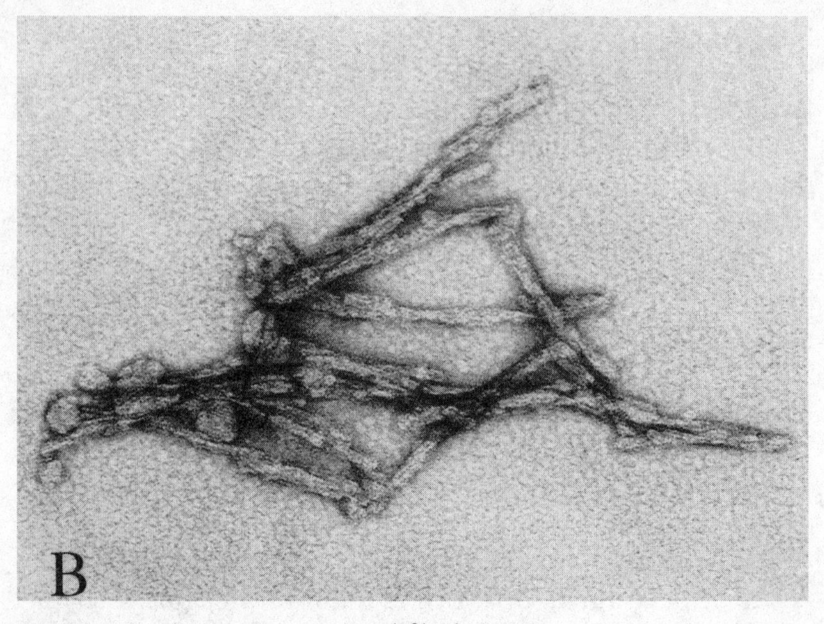

Patricia Merz's scrapie-associated fibrils (SAF).

speculated that her samples were to blame, "perhaps because of their relatively small size and low concentration."

Merz knew she had found something important. She mapped out a strategy for further research and followed it. In her first work with Somerville, she'd studied scrapie-infected mice and hamster brains. Now she collaborated with a husband-and-wife team at Yale University, Drs. Elias and Laura Manuelidis, who supplied her with guinea-pig and human tissue infected with Creutzfeldt-Jakob disease. Merz prepared and studied fifty-six different tissue samples—long months of work in the laboratory and at the EM—and consistently distinguished infected samples from healthy controls solely by the presence or absence of SAF.

Before the spongiform encephalopathies infect the brain they incubate in the spleen, a smooth, purple gland located behind the stomach at the back of the abdominal cavity that fights infection by filtering foreign organisms from the blood. Among the samples the Manuelidises supplied Merz were several dozen of spleen infected with scrapie or CJD. Merz found SAF in the infected spleen samples. That meant her fibrils weren't merely brain debris but something far more significant. By then—around 1980—she and her husband, George, had two small children. She was home one evening washing the dinner dishes when she understood what her SAF might be. "In all the long history of the spongiform encephalopathies," she told me, "no one had ever seen the disease agent. Nobody had any idea what it looked like. I was doing the dishes when it dawned on me that with SAF I might actually be looking at the agent itself. My stomach clenched and I just threw up my dinner."

While Merz was discovering SAF, another determined researcher was closing in on the scrapie agent from another quarter. Biochemist and neurologist Dr. Stanley Prusiner had been a neurology resident at the University of California–San Francisco School of Medicine when he lost a patient to Creutzfeldt-Jakob disease in 1972. The disease intrigued him. "I started reading about scrapie," the Cincinnati native told science reporter Gary Taubes several years ago, "and it became clear that this was a wonderful problem for a chemist. It had been attacked by pathologists, physicians, veterinarians. Those who tried to unravel the chemistry of the disease hadn't taken a very careful approach. . . . I set up a lab here. I got some money from the neurology department, but not a lot."

Prusiner was then twenty-nine years old, a trim, handsome man with a strong jaw, prominent eyebrows and a full head of dark, curly hair. Research that requires large numbers of laboratory animals is expensive. With remarkable chutzpah,

Prusiner applied to the NIH for a major grant for scrapie research. As he remembers it, "They said, 'Who the hell are you?'" He was advised he needed training in virology and experience working with scrapie and his grant application was rejected.

Prusiner's colleagues say they have never known anyone to pursue a Nobel Prize so relentlessly. One example they cite is his decision to study virology at a laboratory in Sweden, whence the coveted prizes originate. After virology training, Prusiner moved to the Rocky Mountain Laboratory to collaborate with Bill Hadlow. Hoping to isolate the scrapie agent, they worked with mice, which produced results faster than working with goats. But waiting out the deaths of successive generations of mice taught Prusiner what his predecessors had already learned, that progress using traditional methods was measurable in successive generations of scrapie researchers. Prusiner and Hadlow worked their way through ten thousand mice before Joe Gibbs cut off funding in 1978. He did so, he says, to push the two scientists to publish their results.

Prusiner returned to UCSF, where he had started. Borrowing a technique the British had developed back in the 1960s, he devised a faster way of measuring scrapie infectivity: he used hamsters that incubated the disease twice as rapidly as mice and he took their measure at the onset of symptoms rather than at death. He estimates that the new system accelerated the work "by a factor of one hundred. Instead of observing sixty animals for a year, we can assay a sample with just four animals in sixty days." Over the next several years, he told Taubes, "we did more experiments on the biochemistry of scrapie than everyone else in the history of scrapie combined."

Sooner or later, all roads in TSE research lead to Carleton Gajdusek. Pat Merz began working with Gajdusek and Gibbs in the late 1970s. Gajdusek continued to divide his time between the United States and New Guinea, and in 1978 and again in 1980, Prusiner pilgrimaged to the Eastern Highlands to add kuru to his quiver. During the 1978 visit, the San Francisco researcher worked up clinical studies on fifteen kuru patients in collaboration with Gajdusek and Mike Alpers. Alpers says Prusiner "wasn't much of a walker. We had to leave him behind on one of our patrols and come back the next day and pick him up." Alpers is one of the few people who has worked with Prusiner who remembers him fondly. The two men subsequently collaborated on a study of cannibalism in hamsters, feeding them scrapie-infected hamster heads and monitoring the resulting infection. Oral transmission of scrapie proved to be "extremely inefficient," requiring a dose of infected tissue one hundred million times greater than did direct inoculation into the brain.

Prusiner's hiking skills hadn't improved by the time of his second visit, Gajdusek recalls. "Prusiner came to visit me in 1980 at Agakamatasa, the village where I lived. We hauled him over the mountain range. He arrived almost dead, and stayed in my bush house a couple of nights. We were in continuous bull sessions for all that time, discussing the future of kuru and CJD and scrapie work."

By then, Gajdusek and Prusiner had both concluded that the scrapie agent contained no nucleic acid. They reached that revolutionary conclusion for the same reasons—because chemicals and processes that destroyed nucleic acid left scrapie fully infective, but chemicals and processes that destroyed proteins reduced or prevented infection—but Gajdusek, a bolder explorer, had come to it earlier. When Prusiner visited New Guinea in 1980, he was in the process of repeating all the classic scrapie-agent challenges using enriched samples that gave less ambiguous results than had earlier studies. In the course of their Agakamatasa discussions, Gajdusek says, "I pointed out to him that I would give the disease agents a proper name when we were sure what their molecular structure was. I made this point repeatedly with him, explaining that it was premature to name them since, although we knew they had no nucleic acid, we were not sure of their biochemical nature. I had not realized that Stan would not give me the prerogative of naming them when the appropriate information was at hand. It was a clever political move on his part to jump the gun."

Possession is nine-tenths of the law. Prusiner went home, finished up his scrapie assaults, decided he knew enough about the biochemistry of the scrapie agent and brashly wrote a paper proposing a name. The American journal *Science* published the paper in April 1982 under the title "Novel Proteinaceous Infectious Particles Cause Scrapie." Prusiner became its sole author after a collaborator who felt it stretched the evidence withdrew from participation. "Six separate and distinct lines of evidence," Prusiner announced, "show that the scrapie agent contains a protein that is required for infectivity." Then, like Jehovah, he blessed the agent with a name:

> In place of such terms as "unconventional virus" or "unusual slow virus-like agent," the term "prion" (pronounced *pree-on*) is suggested. Prions are small *pro*teinaceous *in*fectious particles which are resistant to inactivation by most procedures that modify nucleic acids. The term "prion" underscores the requirement of a protein for infection. . . .

But by jumping the gun, Prusiner had exposed himself to a devastating risk. If the scrapie agent really was an infectious protein devoid of nucleic acid, as he and Gajdusek believed, the name he was proposing would stick. If the agent proved eventually to contain nucleic acid, then it would be just another virus, however slow or unconventional—and Prusiner would be wrong. Wrong guesses don't win Nobels. So the ambitious neurologist added a hedge to his *Science* article. "He hedged much more than I did in those early 1980s," says Gajdusek. To his definition of "prion," Prusiner added a qualification: "current knowledge does not allow exclusion of a small nucleic acid within the interior of the particle." And he went on to discuss two possible models: "a small nucleic acid surrounded by a tightly packed protein coat" or "a protein devoid of nucleic acid, that is, an infectious protein."

Gajdusek says he had given up on nucleic acid "long previously." He was wisely waiting for the evidence to catch up with his hunch. With a Nobel already in his pocket, he could afford to. Evidently Prusiner felt he could not. When British scientists saw the *Science* article, they were understandably hostile to Prusiner's preemption. Alan Dickinson twitted Prusiner in a *Lancet* editorial, invoking the pre-DNA era when "it was generally thought that viruses were proteinaceous, which is now easily seen as the type of half-truth that puts the cart before the horse. . . ." The Edinburgh geneticist reminded Prusiner that the evidence of differing and even competing strains was the crux of the problem: nucleic acid could easily code strain information, but protein had no known mechanism for doing so. Later, Dickinson says, when he met Prusiner, the American had the gall to tell him he didn't believe in strains. Whether Prusiner believed in them or not, he discussed them in his *Science* paper, speculating that multiple genes coding for different proteins might account for strain variations. Which implied, fantastically, that there might be not one prion but a whole phalanx of prions that caused scrapie—"different proteins," as Prusiner phrased it, "with the same biological activities."

Stanley Prusiner's assault on strains particularly irked Alan Dickinson because Dickinson had pioneered strain studies. In 1983, with his colleague G. W. Outram, he had also proposed an ingenious but plausible alternative to both the viral and infectious-protein theories. Plant scientists discovered decades ago that certain diseases were transmitted from plant to plant by short stretches of naked plant DNA. They named these unusual agents viroids. A typical example, Dickinson explained to me in Edinburgh, is a disease of coconuts called kadunkadunk that coconut harvesters inadvertently spread on the spikes they wear to help them shinny up coconut palms. Dickinson and Outram postulated that the TSE agent might be a similar form: a small piece of naked nucleic acid that hijacks a host protein and uses its equipment to reproduce.

Looking for a name, Dickinson remembered the name the great twentieth-century Italian physicist Enrico Fermi had assigned to a small nuclear particle that shared some characteristics with the neutron. Fermi had called his little neutron-like particle a neutrino—a baby neutron. So Dickinson named his hypothetical small, naked virus a virino. He recalls Peter Medawar's quip to define his virino informally as "bad news securely wrapped in somebody else's protein." The virino, if it existed, would supply some of the missing elements in the TSE-agent story. It would explain, for example, why a protein seemed to be required for infection. There's no known mechanism whereby a protein can mutate, but by providing for a transfer of nucleic acid, which *can* mutate, the virino would also explain the existence of strains. Fermi's neutrino was elusive. The Italian physicist postulated it in 1933 to account for a small amount of energy missing after a nuclear reaction. It was such an elusive particle, however—capable of passing cleanly through the earth without leaving a trace—that it was not actually identified until 1956. No one has yet identified a virino.

Prusiner's bravado, however rude, was a form of gambling, a game he would win if his research paid off. Payoffs spilled rapidly from his San Francisco lab in the months and years to come. By December 1982, Prusiner and two of his colleagues could report having purified a protein from scrapie-infected hamster brains that seemed to track with infectivity—the more infective the tissue, the more of the protein they found. A year later they reported that their purified protein, now designated PrP (for *Pr*ion *P*rotein), took the form of "rod-shaped particles" that stained like amyloid. They called the particles amyloid rods. (Later, Prusiner would dub them prion rods.) "Each rod," they wrote, "may contain as many as 1,000 PrP molecules. Our findings raise the possibility that the amyloid plaques observed in transmissible, degenerative neurological diseases might consist of prions."

The rods, of course, were Pat Merz's scrapie-associated fibrils. With purer samples, Prusiner had succeeded in staining them with Congo red. He doggedly denied (and still denies) that Merz's SAF and his prion rods are one and the same, but a paper published by a German group in the same month as his paper on amyloid rods had already established their equivalence. The German group, working with the same line of hamsters that Prusiner favored, had used his purification techniques to concentrate the scrapie agent, had found SAF in the concentrate and had tracked the SAF with infectivity, just as he had tracked his amyloid rods with infectivity. Merz finished the job a few years later by challenging SAF with antibodies to PrP. SAF and amyloid rods reacted identically to the antibody challenge. Which means that the credit for discovering what might be the TSE disease agent properly belongs to Merz, not to Prusiner.

In 1984, Merz, Gajdusek, Gibbs and several colleagues finally got around to publishing Merz's important finding of SAF in mouse spleens—the discovery that had made her lose her dinner. They also reported finding SAF in kuru and CJD brains but not finding SAF in brains damaged by Alzheimer's, Parkinson's and ALS. These investigations justified a bold claim: that SAF were "a specific marker for the 'unconventional' slow virus diseases and may be the [disease] agent" itself. Finding SAF in tissue samples has since become the definitive test for spongiform encephalopathy.

Prusiner in the meantime closed in on the structure of PrP. He demonstrated in 1984 that PrP was at least a *component* of the scrapie agent, if not the agent itself, but was not measurably present in normal brain. That was what would be expected of a disease agent, but Prusiner didn't trust his measuring method enough to push such a claim. He was wise not to. Within a year, he and a vigorous, energetic Swiss researcher, Dr. Charles Weissmann, chemically decoded the PrP molecule and used that protein sequence to locate the corresponding DNA sequence in a gene library. The two men weren't surprised when the PrP gene turned up in the cells of scrapie-infected hamsters—but they were greatly surprised when it turned up in the cells of *normal* hamsters and normal humans as well.

Prusiner and Weissmann next showed that the two forms, the normal and the diseased, had identical DNA codes. How could a normal and a diseased protein share the same specifications? The men looked for other differences. They found only one, but it was important: normal PrP was easily broken down by an enzyme that digests proteins, while abnormal PrP resisted digestion. A protein's shape usually determines its resistance to enzyme activity. But a protein knots itself into a shape only *after* it is assembled as a chain, *after* the DNA that specifies it is copied.

Which seemed to mean that the brain made normal PrP but the normal PrP sometimes changed to a diseased form *after* it was made. That would explain one of the strangest features of the spongiform encephalopathies, the feature that had misled Gajdusek for years into suspecting that kuru was genetic: the fact that the invading disease agent stimulated no immune response whatsoever in its victims. The immune system recognizes and attacks foreign proteins. It leaves its own proteins alone—and PrP, normal or abnormal, was a host protein.

But what caused a potential victim's normal PrP to change to a diseased form? Asking the question makes it clear that the mystery of the nature of the scrapie agent still remained unsolved. One answer could be: a virus or a virino caused the change. That would make the formation of abnormal PrP just another example of amyloidosis, like the amyloid formation that sometimes follows infections such as tuberculosis. A viral cause would also address the strain problem, since mutations in the virus's or virino's DNA could account for the differences in strains. No one had yet found a virus or a virino associated with the TSEs, but not finding one or the other didn't mean it wasn't there.

Prusiner believed prions caused the change: inoculation with abnormal PrP somehow caused normal PrP to reshape itself into the diseased form. No one knew what cells made PrP for, but it was known that cells made PrP continuously. If the PrP changed form, cells would presumably make the diseased form instead of the normal form. That would explain how the amount of infective agent could increase without nucleic acid replication.* But there were at least twenty different strains of scrapie agent; how could there be twenty different kinds of prion? And since many other cells in the body besides nerve cells make PrP, why did prion infection only wreak havoc in nerve cells?

By the late 1980s, everyone in TSE research was frustrated. Despite years of work, there was obviously more to

---

*If this theory sounds like Griffith's Second Way, it probably was. On the other hand, as Carleton Gajdusek likes to say, quoting his mentor Linus Pauling, ideas are the shit of creative brains, shed continuously by all creative people. Credit customarily goes not to an idea's originator but to whoever works out its reality in the world.

learn. Some would continue looking for a virus or virino. Others would try to invent an experiment that would show definitively whether or not the infectious agent was protein alone. Carleton Gajdusek would refine a radical new theory. The general mood of discontent communicated itself around this time to Lewis Thomas, the distinguished American physician who was also a popular essayist. In his best-selling book *Late Night Thoughts on Listening to Mahler's Ninth Symphony,* Thomas declared the scrapie agent "the strangest thing in all biology and, until someone in some laboratory figures out what it is, a candidate for a Modern Wonder."

If the scientific stakes were high, the human stakes would soon go higher. Even as Merz, Prusiner, Gajdusek, Weissmann, Dickinson and a growing crowd of younger researchers attacked the problem with the powerful new tools of molecular biology, a terrifying new form of the disease was insinuating itself into the human food supply, where it would unleash a major epidemic.

Part Three

God in the Guise of a Virus

ELEVEN
Meat Bites Back

Great Britain, 1985–1995

THE PLAGUE THAT STRUCK the cannibal Fore in the Eastern Highlands of New Guinea struck British cattle in April 1985. Dr. Colin Whitaker, a burly English veterinarian with broad hands and a shock of dark hair, attended the first known case. Whitaker lives in Kent, the county of hop fields and dairy farms in southeastern England that includes Canterbury and Dover. He calls the plague of cattle that followed the outbreak "a horror story."

Since they handle them twice a day for milking, farmers know their dairy cows, big females with soft eyes, calm dispositions and udders like downy kettle drums. "It was the twenty-fifth of April 1985," the veterinarian remembers, "when one of my dairy clients phoned up to say he'd got a cow behaving oddly and would I come and have a look at it." Whitaker drove to Plurenden Manor farm outside Ashford, in central Kent, where his client maintained a herd of three hundred milking Holsteins—large-boned black-and-white animals, a valuable dairy breed that originated in northern Germany and was introduced into England commercially in the late nineteenth century. One of the Plurenden Manor Holsteins was sick. "When you approached her," Whitaker recalls, "she would shy away. She was previously a quiet cow and had started becoming aggressive, rather nervous, knocking other cows, bashing other cows and so on and becoming rather dangerous to handle. She would also at the same time become rather uncoordinated. If you shooed her, she would stumble, particularly on the back legs, and go down, and then scrabble along."

Whitaker examined the sick animal and diagnosed ovarian cysts, a common cause of aggressive behavior and nymphomania in cows. He treated the condition successfully, but across the next several weeks the cow's staggering and stumbling worsened. The veterinarian considered other possible conditions, such as a shortage of magnesium in the animal's diet, and treated her for them without improvement. "Eventually she went down," he concludes, "was slaughtered and went to the knacker's yard." ("Knacker's yard" is the traditional English term for a rendering plant, a place where dead animals and diseased carcasses are processed into commercial products such as meat-and-bone meal.)

Seven more cows sickened and went down at Plurenden Manor in the next eighteen months. By 1986, three other herds had been identified with three similar cases each in three counties in far southwestern England—Cornwall, Devon and Somerset. Word spread through the dairy industry. Dairy farmers and private veterinarians like Colin Whitaker alerted the State Veterinary Service of the Ministry of Agriculture, Fisheries and Food (MAFF). Since epidemics typically begin with an index case—a first focus of infection—the distribution of the outbreak in widely separate counties was surprising, because the herds involved were closed, with no contact with other cattle and no importations. How did the disease jump from one end of England to the other?

MAFF's Central Veterinary Laboratory in Surrey, near London, began investigations. The condition resembled no disease of cattle previously known. The lab had difficulty acquiring well-fixed brains from suspect cows, but by late 1986 its electron microscopists were able to demonstrate spongiform damage and the brown stars of astrogliosis. The veterinary scientists conducting the study contacted Pat Merz on Staten Island for advice on how to prepare samples to reveal scrapie-associated fibrils. In the one homogenate of fresh brain available to them at the time, in the summer of 1987, they found SAF. They published their first brief report in the British *Veterinary Record* in October, defining a new disease of cattle they designated "bovine spongiform encephalopathy," BSE, although large amyloid plaques were more characteristic of the disease than spongiform damage. Soon the media would nickname the new condition more sensationally for the nervous, aggressive behavior it provoked in normally peaceful animals, calling it mad cow disease. Caricatures of befuddled, loony black-and-white cows became a staple of British cartoonists. Headline writers recycled Noël Coward's "Mad dogs and Englishmen" into "mad cows and Englishmen" to the point of cliché. But the disease was no laughing matter.

By the end of 1987, BSE had appeared in herds throughout England and Wales but not yet in Scotland, some 420 confirmed cases. More ominously, cases were increasing from month to month enough to predict that the numbers of animals dying would double annually—exponential growth, a chain reaction of fatal disease. Epidemiologists, led by Dr. John Wilesmith of the Central Veterinary Laboratory, had begun traveling in May 1987 collecting detailed information on two hundred cases. They achieved that collection by December and keyed it into a BSE computer database. Then they looked for causes. One important clue was the fact that BSE had broken out more or less simultaneously throughout Britain. *All* the early BSE cases proved to be index cases, the first in their herds, which suggested infection from a common source. What did two hundred cows from all over England and Wales have in common? Not veterinary or agricultural chemicals, the epidemiologists determined. Not genetics: the disease affected a number of different breeds. BSE couldn't have been imported into the herds, because many of them were closed. Scrapie was an obvious possibility, but not many cattle farms kept sheep, so how would the cattle have picked up the infection? Not all herds had contact with wildlife such as deer that might carry disease. Not all herds had been inoculated with the several new vaccines that came available during the life spans of the case animals, ruling out a contaminant like the scrapie that had stowed away aboard Bill Gordon's louping-ill vaccine.

Eliminating these other potential common sources from consideration left one probable focus: food contamination. The fact that BSE occurred much more frequently in dairy herds than in beef cattle was a correlative clue. What foodstuff was fed to dairy cattle but not routinely to beef cattle? One in particular seemed to fit this restriction: supplemental protein in the form of meat-and-bone meal.

Beef cattle are fed grass or hay while they're growing, then fattened with grain. For milk production, on the other hand, a dairy cow requires a fortified diet. A good Holstein dairy cow produces about 1,250 gallons of milk per lactation, some thirty to forty pounds of milk per day, and to make so much protein requires a high-protein diet. Dairy farmers feed their cows high-quality hay and grain, but hay and grain alone are not enough to maintain high-volume milk production. For that, for centuries informally, but systematically and commercially beginning during the Second World War, dairy farmers have boosted their milking cows' rations with a twice-daily portion of protein supplement. Soybean meal makes a nutritious vegetable supplement. Britain unfortunately doesn't grow much soybean. An alternative source of protein supplement in Britain and many other countries, including the United States, is meat-and-bone meal from the knacker's yard: the ground, cooked and dried remains of dead animals, including downer cattle and sheep dead of undiagnosed disease. Cattle (and sheep, pigs and chickens) are thus made cannibals in the interest of cheaper milk and meat.

Beef cattle are also sometimes fed meat-and-bone meal during the finishing phase. Calves are fed meat-and-bone meal to maximize their growth. Dairy cows are slaughtered for meat when their milk production drops after three or four pregnancies, and since their meat is tougher than the meat of young beef animals, most of it is ground into hamburger or chopped for meat pies, a British staple. Cattle brains go into hamburger as well. And because only female cattle give milk, male dairy calves—young dairy bulls—are slaughtered for veal. High-quality animal protein is a relatively rare commodity in the world; not surprisingly, every part of beef and dairy cattle is eaten, recycled back into the next generation of animals or processed into commercial products. "You have just dined," Ralph Waldo Emerson once reminded a New England audience about the philosophical consequences of the human taste for meat, "and however scrupulously the slaughterhouse is concealed in the graceful distance of miles, there is complicity, expensive races—race living at the expense of race. . . ." Today we wouldn't say "race"; we'd say "species"—species living at the expense of species. But sometimes meat bites back.

Once the disease that was killing British cattle was revealed to be a spongiform encephalopathy, contaminated animal protein was an obvious source. Joe Gibbs, Carleton Gajdusek and their colleagues at the NIH had reported successful oral transmission of kuru, CJD and scrapie to spider monkeys by unforced feeding of brain tissue in 1980. Stanley Prusiner and Mike Alpers had shown oral transmission via cannibalism in hamsters in 1985. But meat-and-bone meal had been fed to cattle in Britain and throughout the world for decades without transmitting BSE. Had something changed?

To find out, Wilesmith and his epidemiologists arranged

for three government veterinarians who knew the rendering industry to conduct a survey of all the rendering plants in Britain. To give the veterinarians a starting point for possible change in rendering practices, the epidemiologists needed to know when British cattle first became infected with BSE. They looked at the different ages, at the time the first symptoms appeared, of the two hundred animals they were studying. Triangulating back, they estimated that exposure had begun suddenly and simultaneously in the winter of 1981–1982. That estimate—exposure to BSE in calfhood and an incubation period of four to five years—has stood the test of time.

The three veterinarians carried out their survey in autumn 1988. Rendering plants are the goriest expressions of the recycling spirit, hellish places of steam, blood, grease and stink. There were forty-six such plants in operation in Britain in 1988, of which thirty-nine kept useful records. By chopping, grinding, cooking and dissolving on a Brobdingnagian scale they produced tallow—rendered beef fat—and what the British call greaves. The Oxford English Dictionary defines "greaves" as "the fibrous matter or skin found in animal fat, which forms a sediment on melting and is pressed into cakes to serve as meat for dogs or hogs, fish-bait, etc.; the refuse of tallow; cracklings." In modern practice, greaves were cooked down in steam-jacketed stainless-steel vessels from the various materials delivered to the rendering plants from slaughterhouses, deboning plants, butcher shops and farms: fat trimmings, bones, offal (guts, heads, tails, blood, the "offfall" of slaughtering, butchering and knackering), carcasses from cattle, sheep and pigs, even feathers from poultry—in 1988, some 1.3 million metric tons, about three billion pounds.

Some of the plants processed the greaves further into meat-and-bone meal, which is a scab-colored powder (sometimes pressed into pellets) with a pungent salty, bloody fecal smell.

They used a number of different processing systems, some batch, some continuous. An American system for continuous rendering, which was cheaper than batch rendering, had come to dominate the market; the American system rendered wastes at lower temperatures than the processes it replaced. That was one significant change in meat-and-bone-meal processing which the investigating veterinarians discovered. Another was the abandonment of solvent extraction. Extracted tallow is more valuable than meat-and-bone meal, and reducing the fat content of meat-and-bone meal below five percent extends the material's shelf life, so flammable solvents which dissolved fat had been used for many years to enhance tallow removal. But an explosion at a British chemical plant in 1974 had led to the introduction of tough new standards for solvent use, and rather than investing in expensive new machines, the rendering industry had decided largely to forgo solvent extraction. The proportion of meat-and-bone meal produced using solvent extraction dropped between 1981 and 1982 from fifty percent to only ten percent, while the fat content went up from below five percent to about twelve percent.

Fat protects microorganisms from heat. The epidemiologists concluded that the combination of lowering processing temperature and abandoning solvent extraction had protected the hardy BSE agent from inactivation and thus spread the disease—much as CJD had spread in contaminated human growth hormone or kuru in undercooked human brain. Meat-and-bone meal is generally produced and distributed locally; when the epidemiologists learned that the only two plants that had continued to use solvent extraction were two in Scotland that produced most of that region's meat-and-bone meal, they took the difference for corroboration of their theory: Scotland had been the last region of the country to report BSE outbreaks.

The British government hadn't waited for Wilesmith's epidemiological study to be completed to move to address the epidemic—though it would later be criticized for moving too slowly nonetheless. In the spring of 1988, it had set up a committee chaired by an Oxford zoologist, Sir Richard Southwood, to study the problem and recommend government action. The committee had immediately recommended imposing a ban on feeding ruminant-derived protein to ruminants (meaning in this case cattle and sheep) and in July, the government had complied.

Behind the ban was the assumption that BSE was scrapie, not picked up directly on the farm but passaged to cattle through infected sheep waste incorporated into meat-and-bone meal by inadequate processing. If BSE wasn't scrapie, British veterinary scientists argued, what was it? No such native disease of cattle had been seen before the Plurenden Manor outbreak, at least not officially. In fact, there were other possibilities, and they would eventually be debated among scientists if not officially acknowledged. But the scrapie hypothesis was the Central Veterinary Laboratory's "primary hypothesis," according to Wilesmith. Tragically, assuming that BSE derived from scrapie predisposed British officialdom also to assume that the disease was not likely to spread to humans from BSE-infected meat—because scrapie had not spread to humans despite hundreds of years of eating scrapie-infected lamb and mutton. British officialdom concluded that however costly BSE might be to the cattle industry, the risk of BSE infecting humans was nil. So the first ban the government imposed, the ban on ruminant-derived protein, was intended to protect animals (although in fact it would not be properly policed for years). The ban plainly did *not* protect humans.

A month after the ruminant-derived-protein ban, farmers were barred from marketing cattle obviously ill with BSE. But this restriction did not extend to animals from BSE-infected herds, which might be incubating the disease without yet showing symptoms. Another year would pass before even the parts of these animals thought most likely to carry the infection—"specified bovine offals," they were officially called, meaning brain, spinal cord, spleen, thymus, intestines and tonsils—were excluded from the human food supply.

Nor did the British government pay adequate compensation to farmers and processors for their losses. A tall, ruddy, Cambridge-educated colleague of Alan Dickinson at the Medical Research Council Neuropathogenesis Unit in Edinburgh, Dr. Hugh Fraser, remembers getting angry about the government's attitude—"and I don't get angry very easily," he says:

> The real culprit is the British Treasury. The Treasury could have solved this whole blooming problem by putting in adequate compensation, buying back the contaminated meat-and-bone meal which they said it was illegal to use. When they brought the ban in, they did not offer any financial recompense for those people who were out of pocket on meat-and-bone meal. Some people had spent tens of thousands of pounds on buying it. Not only the farmer but the miller, the chap who's making the protein concentrate. He's in the same position. He's got stuff all over the place. He bought it on the futures market—it's actually on the futures market. If you have got to get rid of meat-and-bone meal, you use it. So the stuff continued to flow through the system. And once it started to flow through the system it continued to flow through the system, because there was no policing of it.

Similarly, the government offered farmers as compensation only fifty percent of market value for their sick animals—giving them inducement to market those animals for food at the earliest sign of BSE infection.

In December 1988, milk from cows obviously ill with BSE was ordered destroyed. Officially confirmed cases of BSE numbered 2,185 in Britain that year, 1,765 more than the year before. The numbers were almost certainly underreported.

Ironically, the triggering event of the BSE epidemic may not have been the feeding of contaminated meat-and-bone meal to cattle. Evidence turned up later that meat-and-bone meal had probably never been processed at a temperature high enough to inactivate the BSE agent. An increased *percentage* of meat-and-bone meal fed to cattle may in fact have triggered the epidemic. "Indeed," reports the science journal *Nature,* "one factor involved in the spread of BSE in the United Kingdom was that inclusion of meat-and-bone meal in animal feeds jumped from 1 to 12 per cent during the 1980s, a shift that resulted from a fall in the value of the pound and a corresponding increase in the cost of soya and fish meal." A blip in world financial markets may thus have begun the leak of BSE into the human food supply.

The Southwood Committee issued its report in February 1989. Three of the four members of the committee were retired from active science and none was expert on the spongiform encephalopathies. They had already recommended destroying infected animals, banning ruminant-based protein and excluding suspect milk. Now they emphasized the assumed link between scrapie and BSE and the improbability of scrapie passing to humans ("Scrapie has been endemic in Great Britain for centuries," their report notes, "without there being any evidence to show an incidence of CJD higher than the international average . . ."). They found "no evidence of maternal or horizontal transmission of BSE," meaning no evidence that it spread from cows to their calves or from cow to cow directly. They endorsed the theory that the disease was spread in contaminated meat-and-bone meal from 1981 until the ruminant-feed ban was put in place in July 1988. They predicted that BSE would continue to infect cattle at the rate of about three hundred fifty to four hundred head per month until 1993 for a cumulative total of seventeen thousand to twenty thousand cases, would decline to a low level by 1996 and would subsequently disappear. The disease had been described as a *new* disease in the scientific paper that named it—which meant its behavior could not be known in advance of observation and experiment—but the Southwood Committee nevertheless confidently predicted that cattle would probably "prove to be a 'dead-end host' for the disease agent," not passaging it further into other animals. (This prediction had already been demonstrated to be wrong; earlier in the report the Committee had noted that "mice inoculated with brain homogenates from two separate BSE cases have, after 10–11 months, shown signs of an encephalopathy . . . indistinguishable from sheep scrapie.") The Committee therefore thought it "most unlikely that BSE will have any implications for human health." Patting MAFF on the back for the speed with which the Ministry had "brought forward regulations," the Committee concluded with remarkable recklessness that "the risk of transmission of BSE to humans appears remote."

A crusading physician and microbiologist at the University of Leeds, Dr. Richard Lacey, would soon publicly condemn the British government's handling of the BSE epidemic. In a 1994 book, *Mad Cow Disease,* he scorns the official response to the Southwood Report:

> No wonder the UK government was delighted. The risk of BSE to man was "remote," it would die out spontaneously all on its own once the feed source had been cut off. No real action was needed. Farming could carry on

much as before. There was to be no slaughter of herds, and no curtailment of cattle movements or breeding. Beef was safe.

When the Southwood Report appeared, reported BSE cases were already up to five hundred per month, higher than the steady-state numbers the committee had predicted. By the end of 1989 they would rise to nine hundred per month. By February 1995, they accumulated to 143,109 confirmed cases, not the mere twenty thousand the committee had predicted.

An environmental health officer, David Statham, later told British television that when MAFF ordered the ban on specified offals late in 1989, "There was no guidance as to how that removal was to take place. We had examples of slaughterhouses where they were splitting heads open with meat cleavers, trying to cut skulls open with chainsaws, trying to suck the brain out through the hole that had been used to stun the animal." A veterinarian, Dr. Marja Hovi, described going to a slaughterhouse "to see the loading of the heads, just to check what the conditions were. I saw one hundred fifty, maybe two hundred heads being thrown into a truck in a heap, where you could see the brain material seeping out through the holes when the heads were upside down, running down onto the heads below and contaminating whatever was beneath. When the heads arrived at the plant, the meat from the cheeks—usually this is poor-quality meat—was trimmed off for use in hamburger, for human consumption." A meat inspector, speaking anonymously, reported that "the specified offals were out of control. All of the material was supposed to be collected up, stained and removed from the building. But what we actually saw was specified offals being mixed up with other material." Statham, the environmental health officer, concluded: "We were being given mixed messages, particularly from the government. We were being given the message that this really wasn't a problem anyway and so this was maybe a bit of window dressing, as opposed to a serious public health matter."

Whether by negligence or by intention, the government was misleading the public into believing that beef was safe to eat. MAFF's chief veterinary officer, a gray, balding bureaucrat named Keith Meldrum, had insisted in public interviews since the beginning of the epidemic that it carried no implications for humans. In January 1989 he explained his reasoning on BBC television. "The evidence on BSE," he instructed the British public learnedly, "is derived mainly from our knowledge of scrapie, and there is no evidence, scientific or otherwise, that scrapie does transmit from sheep or goats to man. Using that as our model, we are fairly confident that BSE does not transmit to man."

The government's position made Alan Dickinson furious. As long ago as 1976, well before the BSE outbreak, Dickinson had written that "we should not assume, at this stage of our knowledge, that scrapie agents are never transmissible to man from infected meat, particularly as we know that some types of cooking would not inactivate the infectivity." Studying the transmission of scrapie to goats, he had found the agent present not only in the tissues excluded by MAFF's later specified-offals ban but also notably in muscle—which is to say, in meat. "The way you train the high-flying administrators, the mandarins," he says today, "can be summed up in one sentence. It is to train people to be at ease with their consciences when they take decisions about things they do not understand. They became experts by reading literature. What is a 'recommendation'? Ah, it's a thing that's written in their final report, where they prenegotiated away whatever wouldn't be acceptable." The administrators were directing the British government's response to BSE, protecting the cattle industry. Dickinson, Hugh Fraser and others who had studied scrapie for decades were shunted aside.

Others raised the alarm as well. Richard Lacey credits a young physician, Tim Holt, and a dietitian, J. Phillips, writing in the *British Medical Journal* in June 1988, with first drawing public attention to the possibility that BSE might be a danger to humans. Chief veterinary officer Keith Meldrum countered with a blanket denial ignorant of years of transmission work at Gajdusek's lab, at Compton and in Edinburgh: "We don't believe that there is any risk from the consumption of this type of material because there is no known association between the animal encephalopathies and the encephalopathies in man." In a May 1989 national radio program, a prominent neurologist announced that he and his family had stopped eating beef and warned people to be careful. "After that," Hugh Fraser recalls, "I and senior colleagues were told that we were not to discuss these issues with the media." By January 1990, with 7,136 cases of BSE officially confirmed in Britain during the previous year, 5,051 more than in 1988, the British press had begun to clamor.

The government responded in February by raising compensation to farmers for sacrificing BSE-infected animals to one hundred percent of their healthy market value. At the same time, it tried to put an optimistic spin on recently reported experiments demonstrating that BSE could be transmitted by inoculation to cattle and by feeding to mice. "These results demonstrate that the disease can be transmitted using unnatural methods of infection," an MAFF press release asserted, "which can only be done experimentally in laboratory conditions and which would never happen in the field." ("What is the point of these unnatural experiments if they are not relevant to the issues?" asks Lacey with exasperation.) The European Union thought otherwise. In March, it restricted the export of cattle from Britain to animals under six months in age, provided that they would be slaughtered before reaching that age—in other words, veal calves. All other British cattle exports (not meat exports) were banned.

The death of a cat in May 1990 caused a national panic in Britain. For a few weeks, beef consumption dropped by a third and London schools temporarily banished British beef from their menus. Max, a neutered five-year-old male Siamese, shared a household in Bristol with four other cats. His owners had noticed in December that Max sometimes seemed to fall asleep standing up. A month later he began staggering. His head tilted permanently to the right. Sudden noise startled him. His skin twitched and the twitching evoked frantic licking and chewing. By May, Max's back legs had failed him, making it difficult for him to urinate or defecate. At that point his owners mercifully had him euthanized. Naturally occurring spongiform encephalopathy had not been reported previously in cats. Chief veterinary officer Keith Meldrum described the case in a letter to the *Veterinary Record,* stressing disingenuously that "there is no evidence that the condition is transmissible nor is there any known connection with the other animal encephalopathies." Publicly, Meldrum was even breezier. "No cause for alarm at all," he said on television on May 10. "This is only one cat death out of seven million cats in the UK." Sixty-two other domestic cats would die in the next four years before the government admitted that contaminated pet food probably killed them.

Richard Lacey, the outspoken Leeds physician, is a ruddy, portly man with a high forehead and thinning white hair who would not look out of place in an eighteenth-century English coffeehouse. He attracted national attention for the first time in regard to BSE after Max's death when he told the editor of London's *Sunday Times* that the disease was still increasing among British cattle, implying that it was spreading through maternal or horizontal transmission, and that all the infected herds should be destroyed. The *Sunday Times* translated Lacey's statement into a headline: "LEADING FOOD SCIENTIST CALLS FOR SLAUGHTER OF 6 MILLION COWS." That blunt proposal summoned a carpet-bombing of reassuring press releases from MAFF. The Minister of Agriculture himself, a red-cheeked young Conservative named John Gummer, assured himself a place in British history by feeding his daughter Cordelia a hamburger before a mob of TV cameras while insisting that British beef was safe.

By the time of Max's death, a number of zoo animals had died of novel spongiform encephalopathies. A nyala (an African antelope) had been diagnosed all the way back in June 1986. Subsequently, oryx, eland, kudu, a puma, a gemsbok, cheetahs and possibly ostriches would be stricken. Some of the zoo animals had been fed rations that incorporated infected meat-and-bone meal, and the biological characteristics of their infections matched those of BSE-infected cattle. But some of the kudu and eland and one of the oryx were born after the introduction of the ruminant-feed ban, and one kudu was the *offspring* of a feed-infected kudu, which meant the disease may have passed from animal to animal in these cases, maternally or horizontally.

Richard Lacey appeared before the British Parliament's Agriculture Committee in June 1990. Although a Conservative member of Parliament who owned a meat-packing company tried to discredit Lacey's testimony, he held his ground and made several important points. One was that meat from BSE-infected animals could not be considered safe even with a ban on specified offals, since lymph vessels and nerves, both known to be infectious, thread through muscle. Another was that there was no evidence linking BSE with scrapie in sheep and in fact every reason to believe that the disease arose in cattle independently, since sheep and cattle had grazed together for generations with no known passage directly between them. In the absence of some test for BSE infection, Lacey argued, the only rational government policy was to slaughter the infected herds. Otherwise, infected animals that were not yet symptomatic would certainly pass—were certainly passing—into the human food supply. Ominously, Lacey translated a statement from a government report into plain English to reveal the real extent of the government's BSE blunder:

> I would like to just read a section from the Tyrrell Report: "Many extensive epidemiological studies around the world have contributed to the current consensus view that scrapie is not causally linked with CJD. It is urgent that the same reassurance can be given about the lack of effect of BSE on human health. The best way of doing this is to monitor all UK cases of CJD over the next two decades." That to me is a report that has something missing, the second sentence does not follow on from the first. . . . I find it unbelievable that a group of independent scientists are saying that we want to reassure the public and the best way of doing it is to measure the incidence of CJD over twenty years. They are proposing the biggest experiment they have ever done to see how many people succumb. I cannot believe this is a view of honorable independent scientists. I believe these reports have been manipulated by other people. I just cannot believe that a scientist will say: *"In order to find out how big the problem is we are going to see how many people die."*

Two British dairy farmers died of CJD in 1993. The government argued that the deaths could be accounted for within the worldwide incidence of sporadic CJD of one case per million people per year. But a freckled, blond, blue-eyed fifteen-year-old, Victoria Rimmer, began coming home from school exhausted in May 1993 and the illness that followed could not be so glibly explained away: CJD is extremely rare in people under thirty years of age. Her grandmother Beryl, who was raising her, said later that Vicky "began losing weight dramatically and I thought she'd got anorexia. I started getting calls from friends and the school asking, 'What is wrong with Vicky?' As the weeks wore on, she became worse. She was falling everywhere, like you see those cattle falling. She kept saying to me, 'What's the matter with me, Mum?' She just couldn't understand what was happening to her. We'd be out in the town and she'd have to sit down on the floor because she felt so terribly tired in herself. Between the end of May and August, she got a terrible pain in her arm and neck—she was in agony, crying all the time. I took her to the doctors eight times but they said there was nothing wrong."

Beryl Rimmer eventually found a doctor willing to admit her granddaughter to a hospital for tests. He thought a measles virus had infiltrated Vicky's nervous system, but he found no virus in her spinal fluid. Next he did a brain biopsy. Three days after the biopsy, Vicky, who "was more than normal," her grandmother says, was "full of life," who loved ballet, slipped into a coma. The biopsy of her brain revealed spongiform encephalopathy. Mrs. Rimmer didn't understand the term and asked the doctor to write it down. He was afraid to do so, she remembers, and told her to look it up in a medical book. Eventually she convinced him. She showed the note to another doctor. He told her "spongiform encephalopathy" was mad cow disease.

A physician investigator from the government CJD-surveillance unit in Edinburgh visited Beryl Rimmer and examined Vicky. Mrs. Rimmer says he warned her to keep quiet about her granddaughter's condition. "Think about the economy," he told her—"think about the Common Market."

In March 1994, a sixteen-year-old schoolgirl, a Muslim born in England of Turkish-Cypriot parents, developed numbness in her face and fingertips and a backache following a fall. By August her speech had become slurred and her balance impaired. By September her memory was failing and she stumbled when she walked. She had never been treated with human growth hormone, never had tissue grafting or neurosurgery. She normally ate lamb but sometimes ate corned beef and hamburgers. Her condition progressively worsened and she was hospitalized. A biopsy revealed spongiform change in her cerebral cortex as well as numerous plaques. The biopsy tissue reacted to antibodies for PrP but not for Alzheimer's amyloid.

Also in 1994, an eighteen-year-old schoolboy was referred to a psychiatrist for depression. He told the psychiatrist that his memory had been failing for the past six months and that he had "gone nutty." A report of the progress of his illness appeared in *Lancet* the following year:

> He subsequently developed visual hallucinations, delusions of reference, and an excessive fear of water and sharp objects, refusing to wash or shave. His parents noted that he had difficulty carrying out simple tasks such as unlocking a door or eating a boiled egg. He was admitted to a psychiatric hospital. . . . His hallucinations continued. . . . His gait deteriorated. . . . For a period of eight years [earlier in his life] he [had] visited his aunt's farm annually and would have drunk unpasteurized milk and been exposed to cows. No cases of BSE have been reported in this herd.

A biopsy tested positive for PrP. The young man died in 1995. Autopsy revealed spongiform change and astrogliosis.

These two deaths (Vicky Rimmer lived on in her coma into 1996) and the deaths of the dairy farmers were not yet sufficient evidence epidemiologically that BSE had begun to infect humans. "The hundred thousand UK farmers with BSE [in their herds]," Carleton Gajdusek wrote me in January 1996, "means at least a million [animal-industry workers] exposed to contaminated meat-and-bone meal and sick BSE cows and this could mean one worker with CJD per annum is to be anticipated." Simply from the one-in-a-million worldwide incidence of sporadic CJD, Gajdusek meant, one in a million animal-industry workers in Britain should show up annually with CJD. "Two cases [among these workers] would be many," Gajdusek continued. "The two cases of adolescents 17 and 19 years old are younger than any previously in the UK and [there are fewer than] 10 [such adolescent cases known] in the world." In fact there were only four cases of adolescent CJD reported in the medical literature up to 1995. Gajdusek's note concluded: "This is *more* cause for concern. Ten adolescents would be an 'epidemic.'"

At the beginning of 1996, besides Vicky Rimmer and the two 1995 cases, seven other young people had already died or were gravely ill.

Seventh Connection . . .

SPECIES	DISEASE	PROGNOSIS
man	kuru (transmits to primates)	fatal
man	Creutzfeldt-Jakob (transmits to primates)	fatal
sheep	scrapie	fatal
mink	transmissible mink encephalopathy	fatal
cattle	**bovine spongiform encephalopathy**	**fatal**

TWELVE
Ice-Nine

Fort Dedrick, San Francisco, Montana, 1985–1995 / Great Britain, 1996

CARLETON GAJDUSEK GREW STOUT across the years. From the wiry five-foot-eight, one-hundred-and-forty-pounder of his early kuru days, when he hiked up and down the sides of mountains on two meals a day and usually arrived in camp well ahead of his patrol line, he enlarged in middle age to a Buddha approaching three hundred pounds. He had always devoted minimal resources to clothing, traveling with a toothbrush and not much more; his size limited his choices further, to plaid lumberjack shirts and sturdy dark pants held up with clip-on suspenders. With silver hair, a broad, corpulent face and shrewd, deep-set eyes, he came to resemble the character actor Charles Durning. He traveled the world continuously, borrowing a winter coat or wrapping a shawl around his shoulders if the weather was cold, investigating rare and puzzling diseases, transfixing audiences with stories of his adventures, and no one could mistake his powerful intelligence.

For nearly thirty years, Gajdusek has argued passionately for a radical interpretation of the cause of the transmissible spongiform encephalopathies, and though final proof is still lacking, there is much evidence for his interpretation. More ominously, no one has yet proved him wrong. It would be better for us all if someone did, but not even the ambitious

Stanley Prusiner has yet devised the definitive experiment that might settle the question one way or the other.

Gajdusek's interpretation begins with the obvious fact that something must carry information if an infectious agent is to reproduce. In all the rest of biology, that something is nucleic acid. But attacking TSE agents with treatments such as radiation that destroy nucleic acid still leaves the agents infectious. Which means to Gajdusek that some other reproductive process must be involved.

If TSE agents didn't use the universal nucleic acid system, how else might they reproduce and multiply? When Gajdusek began looking for other reproduction mechanisms, one came immediately to mind. It was operating on earth even before living organisms appeared, it was universal and it conveyed information reliably from generation to generation. The old and reliable system Gajdusek thinks TSE agents might use to reproduce is crystallization.

Crystals are organized aggregates of molecules—units of identical pattern stacked together in layers. They solidify out of solutions of liquid when conditions are right. Sugar and salt are crystals, the one boiled down from syrup, the other from brine; snowflakes are crystals; ice is crystalline; so are the stalactites and stalagmites that build up crystal by tiny crystal from limestone dissolved in the dripping water of caves. If you blow your moist breath onto a windowpane on a cold winter day it will form a sheet of ice crystals; outside, you may find a similar sheet of dew crystallized on the windshield of your car. Diamonds and most other precious stones are crystals. Many biological materials are also crystalline. Oysters and other mollusks precipitate crystals to make their shells. We make our teeth and bones by precipitating crystals from biological solutions within our bodies. Proteins, "organized aggregates of molecules," are crystalline. Milk curdles by a crystallization process, precipitating the curds of protein that we call cheese.

As a boy, Gajdusek played at crystal precipitation in his attic laboratory. It's a standard chemistry-set trick. Dissolve crystals of photographic film developer—"hypo"—in hot water in a glass beaker until no more will dissolve, then cover the solution with a clean glass plate, turn off the heat and let it cool. Cooler solutions hold fewer dissolved solids than warmer solutions. In the cooling beaker, the hypo should precipitate out in crystals. It doesn't because it doesn't know what shape to take. Instead, the solution cools into a condition known as supersaturation. Uncover it then and drop in a single crystal of dry hypo and an amazing cascade of crystal precipitation will begin. The crystal seed supplies the template—the information—that the solution needs to organize itself.

With crystals, it's the organization that counts. The same pure carbon that crystallizes as diamond will crystallize into commonplace graphite under different templating conditions. A crystal forms to an atomic pattern. The pattern is the information that's passed along from crystal generation to crystal generation. A piece of the previous generation normally supplies the pattern—the seed crystal dropped into the beaker. "Give me eight bricks mortared together at the proper angles," Gajdusek likes to say to illustrate the point, "and I'll build you an octagonal tower." The piece that supplies the pattern—that serves as a nucleus for crystal formation—is called the nucleant or nucleating agent. Given a seed of nucleant for a starter, the crystallization process will go forward on its own, autocatalytically, new crystals patterning off fragments of those previously formed.

Here, then, Gajdusek realized, was a nonbiological mechanism for reproduction and multiplication. Evolution even used it in some biological processes. But it wasn't entirely reliable. Things could go wrong. Crystals of the same material, patterned by different nucleants, could have different physical properties, with catastrophic results. Instead of diamond, you could get greasy graphite, the commonplace stuff of pencil lead. Gajdusek cites the strange case of an industrial "infection" that occurred during the Second World War. Ethylene diamine tartrate (EDT) is a compound widely used as a purifying agent in industrial processes. A plant that was making EDT became infected with a nucleating agent that caused its EDT to precipitate in an abnormal crystalline state. The "infected" form of EDT didn't work. It was junk. "This infection crippled the industry," Gajdusek writes, "and could not be cured."

The plot of Kurt Vonnegut's 1963 novel *Cat's Cradle* derives from an apocalyptic case of nucleant infection—a fictional case, fortunately for us all. A brilliant scientist invents a new form of ice with a crystalline pattern different from ordinary ice and different physical properties:

> "Now suppose," chortled Dr. Breed, enjoying himself, "that . . . the sort of ice we skate upon and put into highballs—what we might call *ice-one*—is only one of several types of ice. Suppose water always froze as *ice-one* on Earth because it never had a seed to teach it how to form *ice-two, ice-three, ice-four*. . . ? And suppose . . . that there were one form, which we will call *ice-nine*—a crystal as hard as this desk—with a melting point of, let us say . . . one-hundred-and-thirty degrees. . . . When [the rain] fell, it would freeze into hard little hobnails of *ice-nine*—and that would be the end of the world!"

Which is what Vonnegut causes to happen in *Cat's Cradle:* a nucleant of ice-nine is accidentally dropped into the ocean and all the water on earth—a supersaturated solution in ice-nine terms—freezes solid, making it impossible for the many biological processes that depend on liquid water to continue, including the circulation of blood in human veins. "There was a sound like that of the gentle closing of a portal as big as the sky, the great door of heaven," Vonnegut's narrator reports. "It was a grand AH-WHOOM. I opened my eyes—and all the sea was *ice-nine*. The moist green earth was a blue-white pearl."

Gajdusek visualized a similar infective process at work in the TSEs. The infective agent was a piece of abnormal PrP amyloid, a subunit of Pat Merz's scrapie-associated fibrils. The SAF themselves were twisted together from linear stacks of these abnormal PrP protein crystals. Kuru plaques were great crystalline knots of SAF that accumulated in the diseased brain. Gajdusek proposed that a nucleant crystal of abnormal PrP was the TSE infectious agent. When it found its way to the sites where PrP was made in the membranes of nerve cells, it displaced normal PrP as a nucleant, teaching the forming protein to pattern itself instead as abnormal PrP. Thereafter, the cell continued to make PrP, but it ended up in the abnormal form rather than the normal. Normal PrP was used by the cell membrane for some unknown function and then cleared away. But abnormal PrP resisted being dissolved or digested. The cell membrane couldn't clear it away. So it accumulated, blocking or poisoning some essential cell process, damaging the cell until it died. Bodily processes then cleaned out the dead nerve cell, producing the characteristic holes of spongiform degeneration. The brown stars of astrogliosis appeared because astroglia are the brain's infection fighters, its counterpart to the body's white blood cells. Just as white-blood-cell count goes up when the body has an infection, so astroglia proliferated in this odd brain infection.

With this ingenious model in mind, Gajdusek stopped calling the TSE infectious agent a slow virus and began calling it an infectious amyloid. That new name generalized the disease beyond its own rare instance. One of the great breakthroughs in Alzheimer's research came in the 1980s as a direct result of the work of Gajdusek and his colleagues on spongiform disease: Alzheimer's was demonstrated to be an amyloidosis caused by the accumulation of a junk protein in the brain, which revealed itself in amyloid plaques. The Alzheimer's protein, called APP (amyloid precursor protein) is different from PrP. It's specified by a different gene located on a different chromosome, and it has never been shown to be infectious. But the TSE disease process and the Alzheimer's disease process appear to be similar: both involve abnormal nucleation followed by the deadly accumulation of amyloid plaque.

Suddenly, then, Gajdusek and his colleagues found themselves looking at two kinds of brain amyloidosis: infectious (the TSEs) and noninfectious (Alzheimer's). Gajdusek began speaking proudly of the opening of "a new era" in microbiology with the mounting evidence that the agents of kuru, CJD, scrapie, BSE and the other spongiform diseases might be "infectious amyloid proteins." The diseases they caused were clearly brain amyloidoses like the many other amyloidoses that plague the human body when it's sufficiently insulted by massive infection. Similarly, said Gajdusek, "a new era has come to neurology as we have realized that both Alzheimer's disease and the normal aging brain are examples of nontransmissible brain amyloidoses."

In the case of the transmissible amyloidoses, Gajdusek said—the TSEs—they had discovered "a new class of microorganisms which demand broadening our understanding of the basic tenets of biology." In Gajdusek's scheme, DNA still made RNA which made protein—he wasn't proposing to overthrow the central dogma of molecular biology—but an abnormal protein intervened along the way to change the developing protein's final form. Here, in short, was a new disease mechanism that had more to do with chemistry than with biology.

Gajdusek's model of abnormal protein nucleation is demonstrably valid as a disease *process*, but final proof is still lacking that it is also valid as the *cause*. Paul Brown, for one, enjoys playing devil's advocate to Gajdusek's grand scheme. "God in the guise of a virus," he once called the TSE agent. "The whole question of the nature of the agent is obviously crucial," he told me, "but a lot of people don't properly divide it into *two* interesting questions. The first question has to do with what we call the pathogenesis. That is to say, once the process has been *started,* what is the mechanism by which it continues? How is it self-sustaining? How does it wind up as amyloid that you can see? But in point of fact, the more interesting problem is *how does it begin in the first place?*" No one any longer doubted that an abnormal form of PrP was responsible for the damage of spongiform disease. But many continued to doubt that the agent of transmission was an infectious protein. It might still, after all the decades of work, be a conventional virus, or a virino.

Connecting the TSEs with Alzheimer's, generalizing the problem from a group of rare diseases to the most common dementia that afflicts humankind, brought many new researchers and much new money into the field. Since the mid-1980s, a new generation of molecular biologists has gone to work on the TSEs. Polymer chemists knowledgeable about crystallization processes have joined the field (polymers are molecules built up of smaller repeating molecules bonded together; many plastics, but also proteins and nucleic acids, are polymers). Both kinds of scientists, as well as microbiologists and research physicians, are trying to make the breakthrough discovery that would pin the disease agent down once and for all. Their experiments have been elegant in their ingenuity.

Once it was clear that PrP was a host protein—was made by nerve cells, that is, not by an invading virus—it became possible to work back from the protein to the RNA that specified it and from the RNA back to the DNA. The DNA sequences—the genes—for mouse PrP, hamster PrP and human

PrP were all decoded in 1986 and found to be similar, which was evidence at the molecular level for Gajdusek's insight that all the TSEs were basically the same disease. These DNA sequences then served as tools for probing the differences among the various forms of the disease.

Dr. Byron Caughey is a sturdy young biochemist who works with a team led by immunologist Dr. Bruce Chesebro at the Rocky Mountain Laboratories* in Hamilton, Montana, Bill Hadlow's home base. Before Caughey came to Hamilton, Chesebro had decoded the gene for mouse PrP (Prusiner's group did hamsters and humans) and had shown that the gene was identical for both normal and abnormal mouse PrP—arguing in favor of a change of pattern as the difference between the two. In 1988, for the first time, Caughey made abnormal PrP in a test tube. He cloned PrP DNA from a scrapie-infected mouse brain and inserted the gene sequence into mouse cancer cells growing in culture. The cells began making PrP, and because only the gene had been transferred, Caughey could be certain the test-tube PrP wasn't contaminated with any virus. When enough of the altered cells had accumulated, Caughey homogenized them and injected some of the material into healthy mice. The experiment was a first attempt at what might be a definitive demonstration: making abnormal PrP artificially in a test tube, free of any possible contamination, and showing that it can produce infection in scrapie-free animals. Unfortunately, Caughey's mice never came down with scrapie. Which didn't prove the opposite, that the protein isn't infectious: something might have been missing from the mix.

Creutzfeldt-Jakob disease sometimes runs in families. Within such families, an individual's risk of contracting CJD might be one in a thousand rather than the worldwide one-in-a-million risk of sporadic CJD. Such familial jeopardy im-

---

*The NIH institution pluralized its name when it expanded.

plied that genetic differences might determine susceptibility to the disease. But diseases that run in families aren't necessarily genetic, as Gajdusek learned with kuru. A rare variant of CJD that progresses more slowly to death, the formidably named Gerstmann-Sträussler-Scheinker syndrome (GSS), is specifically familial. In 1989, Stanley Prusiner's group, in work led by a brilliant Chinese-American researcher, Dr. Karen Hsiao, identified a mutation in the PrP DNA sequences of two GSS families, one American, one British, that did not show up in a hundred normal people chosen for comparison or in fifteen people with sporadic CJD. The difference indicated that GSS at least was inherited. More spectacularly, it revealed that the mutations determined the course of the rare disease: assuming that the two families were unrelated, the mutations must have arisen independently; they were identical, and they caused identical illnesses. "The new results tip the scales in favor of the 'protein only' hypothesis," Swiss researcher Charles Weissmann wrote in *Nature* that year. They did so, he argued, because they seemed to show that a mutation could "greatly increase the probability of spontaneous conversion of [normal] PrP^{C} to [a diseased form,] PrP^{GSS}, which would then behave like a transmissible agent." Nevertheless, Weissmann added, the results "do not exclude the possibility that a nucleic acid is part of the prion"—because a TSE virus or virino might be widespread in the population but normally latent, and a mutation such as the GSS mutation might reduce resistance enough to allow family members to become infected.

One by one, into the decade of the 1990s, at Gajdusek's laboratory and elsewhere, researchers identified mutations specific to sixteen different familial forms of CJD and GSS. A mutation in an Italian family's genome produced a horrible variant of CJD, called fatal familial insomnia, which blocked sleep in afflicted family members until they hallucinated, lapsed into a coma and eventually died. On the other hand, sporadic CJD cases—the kind that seem to occur randomly throughout the world—usually lacked mutations. So did kuru, reinforcing Gajdusek's conviction that kuru began with a single sporadic case of CJD which was then spread through the Fore by cannibalism, just as BSE was being spread cannibalistically in cattle in infected meat-and-bone meal.

Gajdusek thought this abundant evidence of genetic control of neuropathology supported his protein-crystallization theory. Sporadic CJD, he believed, resulted from the spontaneous conversion within a victim's brain of a normal PrP molecule to the abnormal form that caused disease. The familial mutations, Gajdusek proposed, lowered the barrier to such accidental conversion. "Thus," he wrote in 1996, "with these mutations, this ordinarily rare event becomes a . . . dominant inherited trait." But Weissmann's qualification still remained to be refuted: the mutations might simply allow easier entry to a lurking virus.

Stanley Prusiner was busy during these years. He raised millions of research dollars for his San Francisco team and guided it in what Gajdusek appraises as "first-rate scientific work." In 1989, Prusiner reported developing a line of mice into which he had inserted a hamster PrP gene. The transgenic mice made hamster PrP as well as their own mouse PrP. Challenged with hamster scrapie, which won't infect normal mice, the transgenic mice sickened after an incubation period typical of the *hamster* disease. Prusiner next challenged another group of his transgenic mice with mouse scrapie. With both hamster and mouse PrP genes but with only mouse scrapie challenging them, the mice sickened at *mouse*-scrapie incubation times. "Our findings," Prusiner concluded, "argue that the PrP gene modulates scrapie susceptibility, incubation times and neuropathology." The findings were further confirmation of what the familial-genetics work was showing: that the TSEs were somehow both genetically controlled and transmissible. Such multiplicity was hard to square with the notion of a virus, although Alan Dickinson's virino theory could accommodate it.

Even more remarkable was work Prusiner carried out with Charles Weissmann in 1992 and 1993. The San Francisco and Zurich teams first knocked out the PrP gene from a line of mice, leaving the mice with no way to make PrP protein. "Surprisingly," write the two scientists, "[these knockout mice] develop and behave normally . . . and no immunological defects are apparent." (Mice that developed and behaved normally without a PrP gene raised the interesting question of what purpose the protein served in the body; it was common to many different animals, so it presumably had a valuable function, or evolution wouldn't have conserved it. The question has not yet been answered.) Weissmann's team followed up this surprising result by challenging the knockout mice with mouse scrapie. Lacking a PrP gene, the mice resisted the infection and continued in good health. When Weissmann then inserted a *hamster* PrP gene into his knockout mice and challenged them with both hamster and mouse scrapie, they sickened with *hamster* scrapie but not with mouse. At very least, these experiments showed that PrP was required for scrapie susceptibility.

While Prusiner was accomplishing this important work of his own, he also invaded and colonized the work of others in his apparent pursuit of a Nobel Prize. Gary Taubes, the science journalist, exposed Prusiner's manipulations in late 1986 in a *Discover* magazine article called "The Game of the Name Is Fame, but Is It Science?" The article focused on Prusiner's coinage of the name "prion," his limelighting and his efforts to obscure the fact that no one had yet definitively excluded a nucleic acid requirement for TSE infection. It also exposed Prusiner's manipulations to suppress work Paul Brown had done.

"Taubes started on the West Coast," Brown remembers. "He told me he started with the idea that Prusiner was right and his critics were just expressing sour grapes—he really was pro-Prusiner. He talked with Prusiner a bit and then he talked with some other people. Little by little over a period of months he begin to adjust his glasses and by the time he got to me he was pretty neutral—he was what he should have been at the start. In the course of his discussion with me, I mentioned that I really didn't like Prusiner, had never liked him since I had come to suspect that Prusiner had stone-walled a couple of papers that I'd written. In fact, I knew he had just from the three-page, single-spaced critique he wrote."

Reports of experiments submitted to scientific journals are reviewed by peers—other scientists working in the same field whom editors believe to be knowledgeable—before being accepted for publication. If the peer reviewers' critiques raise questions, editors often return the reports for further work. "Which means," Brown continues, "that you have to do another year or two of experiments before you get something off the ground. That's what I told Taubes. He picked up on it and he went up to the *New England Journal of Medicine,* sat down and talked with the editor. The editor confirmed that what I'd suspected had been true and confirmed further that in the interim Prusiner had had the audacity—the stupidity, actually—to send the journal a paper on the same subject. Now *that* I didn't know, although in retrospect nothing Prusiner does in this line surprises me. It was extraordinary that Taubes got that out of the editor of the *New England Journal of Medicine.* Well, that was a very dark moment for Stan. Because there it was, it was true, and there was nothing he could do about it. He was caught with his hand in the cookie jar." The journal had rejected Prusiner's paper, but the episode left a bad taste; Gajdusek had to push to get Brown's work published.

Prusiner might have learned from Taubes's exposure to go and sin no more. Instead, he decided to stop talking to the media. (He refused to be interviewed for this book.) Because priority—who first makes a discovery—is the accepted measure of achievement in science, he continued to try to claim priority for work other people had already done. That was the offensive he had deployed against Pat Merz and SAF. In 1968, Alan Dickinson, Hugh Fraser and their colleagues at the MRC Neuropathogenesis Unit in Edinburgh had identified a gene in mice that controlled the survival time and neuropathology of various strains of mouse scrapie. They named the gene *Sinc,* which was short for "*S*crapie *inc*ubation." Mice with one form of the gene got scrapie after a short incubation period; mice with another form of the gene got scrapie after a long incubation period. In the early 1980s, Prusiner approached this problem from the perspective of molecular biology and came up with a gene controlling scrapie incubation in mice, short and long. In the report he and his colleagues published in 1986, he proposed calling the gene *Prn-i* (for *Prion incubation*) and showed that it was tightly linked with the PrP gene. Deep in the report, he rejected *Sinc* disdainfully: "The chromosomal location of *Sinc* has not been determined and no linked marker loci have been described." It was little less than outright theft of Dickinson and Fraser's priority.

The two scientists were predictably livid at Prusiner's poaching. They went to work matching up *Sinc* with *Prn-i.* Within a year they were able to report that "the *Sinc* gene and the gene coding for PrP [Prusiner's gene, that is—*Prn-i*] are linked, and could even be the same gene." Gajdusek evaluates the rival claims thus: "Dickinson did all the basic work. Stan only made minor corrections." Gajdusek has seen enough personal conflict in science to tolerate Prusiner's manipulations—after all, he already has his Nobel. "I find humorless Stan rather amusing," he wrote me, "but I fear many of my younger colleagues are less condescending and more bitter."

Even Prusiner's prion theory is essentially a variant of Gajdusek's theory of protein crystallization (and both were preceded, as theories, by British mathematician J. F. Griffith's work back in 1967). By 1996, both versions were ensconced side by side in the bible of virological science, *Fields Virology.* The third edition of this medical textbook and reference carries a long review by Gajdusek of infectious amyloids and a long review by Prusiner of prion diseases. To balance these different versions, the editor commissioned Bruce Chesebro of the Rocky Mountain Laboratories to write an introduction. "The chapters that follow present a detailed analysis of these diseases from two individual perspectives," Chesebro explains of the Gajdusek and Prusiner reviews. "Although they disagree in terminology, they both lean in the direction of assuming that the TSE agent is an abnormal protein." He goes on to remind readers that the involvement of a viral agent had not yet been disproved. I asked Gajdusek about this unusual side-by-side display of scientific ego. He responded with an extraordinary antecedent, two famous eighteenth-century scientists who independently invented a powerful system of mathematics but assigned it different notations, only one of which caught on: "I am in good company for losing the semantic battle, as did Isaac Newton, who lost completely to Gottfried Wilhelm von Leibniz in his fight for the symbolism of the calculus." Prusiner continues to drill his terminology untiringly; his most recent book, edited in 1996, is called *Prions Prions Prions.*

More serious, to those who still remember that a virus or a virino might be involved in TSE, is the extent to which Prusiner's prion theory (Gajdusek's infectious amyloid theory) has stifled debate. "There are very few people now looking for anything," Dr. Janet Fraser, Hugh Fraser's wife and colleague, observes caustically. "People accept the prion, in spite of the lack of evidence and the fact that Stan has only recently acknowledged the existence of strains. It's accepted as gospel, it's taught in all the textbooks. I mean, we have it because it's quoted to us. It's all sorted out, so we can all go home."

Evidence continues to accumulate that the TSE agent may be a misfolded protein. In 1994, Byron Caughey and an MIT chemist, Peter Lansbury, found a way to mix normal PrP with abnormal scrapie PrP in a purified cell-free solution in a test tube. A small virus couldn't multiply in a cell-free environment, because such viruses are parasites that reproduce by hijacking the gene machinery inside cells. Some of the normal PrP in the test tube converted to the abnormal form. Caughey and Lansbury had to use so much abnormal PrP to drive the reaction, however, that there was no way they could pick enough of the converted PrP out of the mixture to test it for infectivity. If they had been able to do so, and if the converted protein had been infectious, they would have had what they call "the ultimate proof of the 'protein-only' hypothesis."

Using this new cell-free system, Richard Bessen, a young researcher working with Caughey and Lansbury, mixed normal PrP with two *different* strains of abnormal scrapie PrP and succeeded in converting some of the normal protein to *each* of the abnormal forms, a major breakthrough reported in 1995. "That's just the sort of thing that Carleton Gajdusek would predict," Caughey told me—"that the two strains have different structures or conformations that can somehow seed a similar change in the normal PrP that they recruit."

Caughey and Lansbury reviewed more evidence that Gajdusek's crystallization theory might be correct in a 1995 paper that paid homage to Kurt Vonnegut's prescient science fiction. The paper was titled "The Chemistry of Scrapie Infection: Implications of the 'Ice 9' Metaphor." The two scientists showed in diagrams how a seed crystal of TSE could serve as a nucleant for abnormal PrP formation, teaching the normal protein to fold up in a different pattern. They pointed out that this mechanism could account for strains: "Strains of [abnormal scrapie PrP] may represent alternately packed, ordered PrP aggregates analogous to the various forms of ice that inspired Vonnegut's ice-nine. . . . Propagation of the scrapie agent strains resembles the seeding of a crystallization." A rare accidental misfolding would account for sporadic CJD; seeding—transmission of the infectious protein—would speed the process up. "The best-known example of . . . nucleation," they noted, "is cloud seeding, which was the field of Vonnegut's brother."

I've known Kurt Vonnegut since 1965, when I interviewed him for the *Paris Review.* After hearing Gajdusek invoke ice-nine and reading Caughey and Lansbury's paper, I reread *Cat's Cradle,* called Kurt and asked him how he happened upon the idea of an end-of-the-world variant form of ice. He told me his brother had been a physical chemist at MIT and had done pioneer work on cloud seeding—making rain by pumping crystals of silver iodide into clouds to supply nucleants around which raindrops could coalesce. I reminded him of the EDT factory that turned to junk and he said, "That too." Then I told him about the TSEs and the possibility that they infected like ice-nine by crystal nucleation.

Vonnegut didn't miss a beat. He said: "Wouldn't you know."

Dr. James Ironside is a pathologist with the British National CJD Surveillance Unit in Edinburgh. In September 1995, studying a brain cross section from the teenage boy who had recently died of Creutzfeldt-Jakob disease, Ironside found amyloid plaques so large that they looked under the microscope like chrysanthemum blooms. They were not confined to the cerebellum, the pathologist determined, but spread throughout the brain. They stained for PrP. Unlike the smaller plaques of ordinary CJD, these florid plaques were surrounded by a zone of spongiform change—a destructive halo of holes.

Ironside had never seen this unusual pattern of damage before, but he knew that sporadic CJD pathology varied widely from case to case. He was startled, then, when another teenage case turned up almost immediately with identical pathology. He alerted the director of the Surveillance Unit, Dr. Robert Will. Will mobilized the Unit's staff. Staff members quickly turned up six more suspect cases in young people. At first Will and Ironside thought the youthfulness of the victims might be the reason for the similarities in their pathology. When the physicians checked the medical literature, however, they learned that the few rare cases of CJD in people under thirty in Britain and Europe showed no such florid plaques widespread in the brain. Late in 1995, Surveillance Unit staff began traveling the country interviewing the victims' families to rule out familial CJD or iatrogenic CJD from growth hormone or surgery.

By the end of February 1996, Will and Ironside knew they had an epidemiological cluster: eight cases of CJD in young people that all showed a new neuropathological profile of florid plaques, early loss of coordination and late dementia. Will arranged to report the ominous new findings to the Spongiform Encephalopathy Advisory Committee (SEAC), a group of scientists and physicians appointed to advise the British government on BSE. A SEAC meeting was set for March 8. Ironside opened the meeting with slides illustrating the unusual pathology. The SEAC chairman, Dr. John Pattison, remembers the moment vividly: "Before he said anything, we could see what it was. It was dramatically different." Another SEAC member, Dr. Jeffrey Almond, recalls near-panic. "The atmosphere became genuinely quite tense.

Some of us were genuinely afraid of what we were hearing. We were afraid that this really maybe indicated a transmission [of BSE] to humans."

Pattison scheduled another meeting of SEAC to discuss the findings further on March 16. By then, Will and Ironside had a ninth case. The committee considered a full range of actions, from leaving the present bans in place to destroying the entire British cattle herd, some ten million animals.

SEAC reconvened with senior government officials on Tuesday, March 19, at the Ministry of Health offices in Whitehall. Will and Ironside now had a tenth case to report. All ten showed the same unique pathology. All had what Gajdusek would later identify as kuru plaques: the florid plaques may not have been seen in Britain and Europe, but they were diagnostic signs of kuru in Papua New Guinea. Some SEAC members were in Paris attending a conference of scientists working on BSE. Stanley Prusiner was there, as were Paul

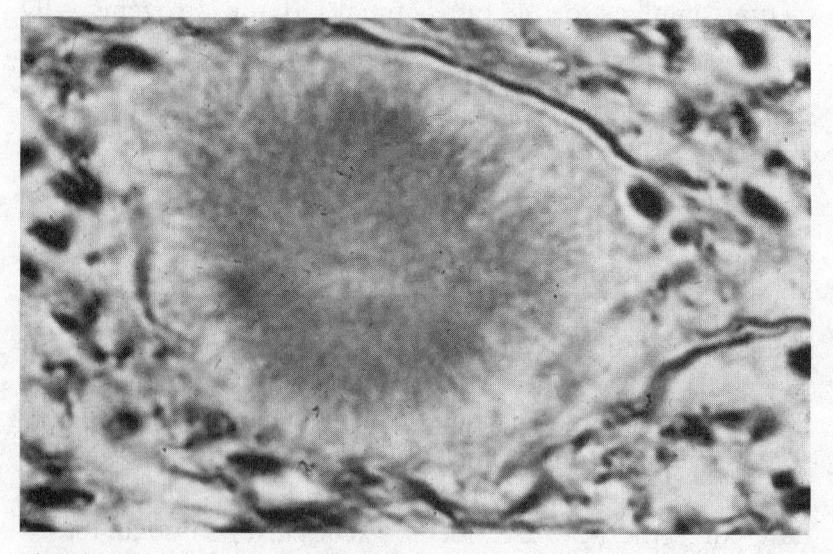

Florid amyloid plaque (kuru plaque).

Brown, Joe Gibbs, Jan Fraser. The SEAC scientists had to be called out of meetings to participate in the London discussions on an open telephone line. The debate went on until midnight. SEAC finalized its recommendations the next morning. They included destroying all British cattle over thirty months of age.

The language of medical reports is impersonal. The ten variant CJD "cases" were human beings, and those who loved them grieved their loss. One was blue-eyed Vicky Rimmer, then eighteen and still barely clinging to life. ("She's blind," her grandmother would anguish. "She can't move. She can't swallow. It's just living hell, seeing her every day.")

Another was the student who reported having "gone nutty." His name was Stephen Churchill. "About four months before he died," his mother told interviewers, "he started to stagger and it brought back memories of seeing the cows on the news. I mentioned it to my husband, could it be this disease that has to do with mad cows. He said, 'No, it can't be, it's too ridiculous.' So I dismissed it."

Another was a thirty-eight-year-old housewife, Jean Wake, who had once worked as a meat chopper in a pie factory. After she died, in 1995, her mother had written Prime Minister John Major, asking if her death could be related to mad cow disease. Major had replied, "I should make it clear that humans do not get 'mad cow disease.'"

Another was a twenty-nine-year-old mother of two, Michelle Bowen, who had worked in a butcher's shop. Comatose three weeks before she died, she gave birth to her second child, a son, named Anthony after his father. "She didn't know what day it was," her husband said bitterly of her delivery, "let alone know that she'd given birth. I want the world to know this and see her picture. My wife is dead. Now we have to wait and see if Tony will suffer the same."

Another was Maurice Callaghan, thirty, of Belfast. Someone at the cemetery where he was to be buried panicked and issued the gravediggers protective clothing and rubber gloves.

The seventeen-year-old Muslim girl was not included among the ten, and four of the cases have not been identified beyond their ages and gender: three women, two aged twenty-eight, one aged twenty-nine; one man, aged thirty-one, like Vicky Rimmer not then deceased.

The last case to be included was Peter Hall, a twenty-year-old student, who was fond of hamburgers as a young teenager but who had become a vegetarian. He died in February 1996. "It was the cruelest thing to witness," his mother would say of his illness, "like babyhood developing in reverse." His father remembered: "When BSE was first discovered I gave up beef because I expected something like this might develop. But not in my wildest nightmares did I think it would strike down someone in our own family."

Robert Will would tell a London newspaper of these ten cases that "their brain tissue displayed a distinctive disease pattern closer to the damage inflicted on a cow's brain by BSE than the damage normal CJD inflicts on humans."

At the SEAC session on March 19, senior members of the British Cabinet tried to suppress any announcement of the new variant form of CJD, arguing that the scientists might be wrong. The Secretary of State for Health, Stephen Dorrell, insisted that the public had to be told. Wednesday, March 20, speaking in the House of Commons, Dorrell informed a stunned nation that BSE had probably spread to humans from eating beef.

Wouldn't you know.

THIRTEEN
It's Kuru and Nothing but Kuru

Stetsonville, Wisconsin, 1985 / Great Britain and the World, 1996–20??

STEPHEN DORRELL'S ANNOUNCEMENT in Parliament of ten cases of Creutzfeldt-Jakob disease attributed to eating infected beef produced world-war-scale headlines throughout Britain and Europe. More British schools banned beef from their cafeterias. Beef sales plummeted until supermarket chains cut prices by fifty percent, after which three out of four customers began buying beef again. ("I love the taste," a dignified London matron told a television news crew. "I just buy the more expensive cuts.") The European Union (EU) blocked imports from Britain not only of meat but also of many beef by-products, including, at least temporarily, such gelatin- and tallow-bonded goods as candies, cookies, lipstick and cough medicine. Many EU decisions require unanimous votes, and John Major's Conservative government retaliated against the beef ban by refusing British agreement across the board. The stalemate persisted for weeks.

I flew to London at the beginning of April 1996, within days of Dorrell's announcement, and was startled to see the words "Creutzfeldt-Jakob disease" in announcement cards in the windows of McDonald's explaining why the fast-food chain was no longer serving British beef. McDonald's decision to switch to Dutch beef at its 660 outlets would cost the

British cattle industry $37.5 million a year, but it was not clear how much longer European beef might be counted safe. Following Dorrell's announcement, the French public-health service revealed that a twenty-seven-year-old Frenchman from Lyons had died in January with lesions identical to those of the ten British variant CJD cases. Until the EU ban, France had imported eighty-eight thousand tons of British beef annually as well as livestock. British meat-and-bone meal had flowed liberally to Europe until 1988, and there was evidence that British suppliers had illegally dumped hundreds of thousands of tons of contaminated product after that. ("I badgered our chief veterinary officer," a British government veterinarian confessed, "saying that having identified a 'poisoned food' it was immoral to export it. But I was firmly put in my place, and told that it was up to the importing countries to put in place all the guarantees needed.")

Shipments to Europe of purebred breeding stock alone totaled 57,900 head between 1985 and 1990. The London *Sunday Telegraph* discovered that more than a hundred thousand veal calves exported to France from Britain in 1995 never went to slaughter. Instead, the newspaper reported, "they were illegally absorbed into herds in France, Italy, Spain and Holland, reared to maturity and are now being sold for beef." So much for Dutch beef. MAFF responded by pointing out acidly that companies like McDonald's which were importing foreign beef might be getting British goods after all.

These legal and illegal exports of animals and meat-and-bone meal made European BSE statistics suspect. All the other EU states together had reported only about four hundred cases of BSE since the British outbreak; statisticians argued, based on the British experience, that the number ought to be above two thousand. Many European countries slaughtered an entire herd with little or no compensation when a single BSE case appeared, discouraging farmers from reporting. "They have something in France called 'tractor' disease," Hugh Fraser told me wryly when I saw him in Edinburgh. "When they get a cow with BSE they buy a tractor and dig a hole and bury it." One expert countered that the low levels of BSE in Europe indicated the contaminated British meat-and-bone meal might have been fed to pigs and poultry. A Dutch veterinarian wise to the ways of the Amsterdam waterfront hinted that some was probably also rebagged and reimported into Britain. Accepting the inevitable, the Federation of Veterinarians of Europe acknowledged in June that BSE had become not just a British but a European problem. Processing protocols in European rendering plants, a British study found, were no more likely than British methods to inactivate the BSE agent.

Nor was BSE-infected beef the only threat to human health. SEAC acknowledged shortly after Dorrell's announcement that it was concerned as well about BSE contamination of lamb and mutton. Natural scrapie was not known to jump the species barrier and infect humans. But sheep as well as cattle had been fed meat-and-bone meal contaminated with BSE-infected cattle remains. In laboratory experiments, the BSE agent had proved capable of infecting sheep as well as cattle, and when such infected sheep tissue was fed to mice, the mice developed lesions typical of BSE rather than scrapie. If beef infected with BSE caused variant CJD in humans, SEAC asked, why would lamb and mutton not do so as well? BSE might also spread horizontally and maternally in sheep, as scrapie did, making its eradication impossible. By July, under pressure from the EU, MAFF felt compelled to add sheep, goat and deer brains, spinal columns and spleens to its ban on specified offals. British Minister of Agriculture Douglas Hogg advised Parliament that the risk of infection was merely theoretical, but he cited an experimental transmission of BSE to one out of six sheep fed BSE-infected brain.

The net hauled up grislier catch. The Swiss government confirmed in April that two clinics in Zurich had been disposing of human placentas—presumably from abortions—by delivering them to a knacker's yard, where they were mixed with animal remains and incorporated into meat-and-bone meal fed ultimately to Swiss pigs and chickens. Hundreds of placentas were involved. The Zurich canton veterinarian pledged to stop the practice immediately and to consider prosecutions.

While I was in London, my daughter Kate, a molecular biologist, called me from San Diego. She had seen Carleton Gajdusek on television and wanted me to know that he had been arrested. (By the time I returned to the United States, Gajdusek's colleagues had raised his $350,000 bond by pledging the equity in their homes. One of the many young people he had brought to the U.S. over the years and impoverished himself to educate, a twenty-three-year-old Micronesian, had accused Gajdusek of forcibly molesting him when he was a teenager. Since the young man had been a champion high-school athlete, those who knew him found his accusations hard to credit. There had also been federal charges, based on anonymous accusations, of fiscal irregularities. Those were investigated and found baseless, but Gajdusek eventually negotiated a plea bargain on the sex charges that involved prison time. However troubling his personal life, his authority as a scientist was never in doubt.)

The Swiss had officially acknowledged 233 cases of BSE in their herds, forty in 1996 alone, the largest number of cases outside Britain (Switzerland is not a member of the European Union). When the Swiss government announced, later in the year, that to restore consumer faith in Swiss beef it would destroy some 230,000 cattle born before it outlawed meat-and-bone meal in November 1990, it undermined its credibility by announcing as well that the remains would be processed into meat-and-bone meal and fed to pigs. Evidently the pigs would be exported: the two major Swiss supermarket chains advertised at the same time that they would no longer sell meat from any species raised on feed containing animal protein.

French and British researchers reported in June that they had successfully infected rhesus monkeys with BSE and that the resulting lesions looked like the new-variant CJD (vCJD), with florid kuru plaques haloed with holes. "It's the first experimental argument," the French researchers told the media—"and a very strong one—in favor of a link" between the two diseases. BSE transmission to macaques, reported the same month in *Nature,* gave further evidence of a link: ". . . The pathological 'signature' of the BSE agent in . . . macaques is identical to that of vCJD in humans." No florid kuru plaques were seen in two macaques inoculated at the same time with sporadic CJD. Interpreting these results, a Swiss neuropathologist found "unsettling" the fact that the modest amounts of infected tissue inoculated directly into the brains of the vCJD macaques were "well within the range of brain tissue present in commercial food products for human consumption until a few years ago." We can hope, the neuropathologist observed, "that the oral route of administration will be considerably less efficient." But the six sheep Douglas Hogg had mentioned had each been fed only five hundred milligrams of brain extract—about one-fiftieth of an ounce—and the fact that so small an oral dose had infected one with BSE did not encourage optimism.

Neither did British research, reported in August, confirming that BSE passed from cow to calf. The seven-year study found a maternal-transmission rate of only ten percent. Douglas Hogg cautioned the media to "keep this information in perspective." There was, the Minister of Agriculture said, "no case for changing recommendations in relation to milk, meat, blood or any other product which is currently permitted." A group of scientists from Oxford, MAFF's Central Veterinary Laboratory, Compton and the University of Colorado reviewed the BSE epidemic in the light of the new study and concluded that BSE infection of cattle was "well past its peak, and seems to be in a phase of rapid decline. New infections from contaminated feed are predicted to be close to zero by the end of 1994,* with all new cases of infection arising from maternal transmission. However, the numbers are small and this route of infection by itself cannot sustain the epidemic." The scientists predicted that "the epidemic is likely to fade close to extinction by the year 2001 [even] in the absence of culling."

The British government had slaughtered 257,000 cattle over thirty months of age since the announcement in March 1996 of a link between BSE and vCJD in humans; a secondary program to cull the most afflicted herds of some 127,000 younger animals was scheduled to begin shortly as well. Licensed disposal plants had been unable to keep up, farmers had been forced to waste valuable feed to maintain animals that would only be destroyed and Britain was running out of cold-storage space for the backup of unprocessed remains. Nor did the slaughter have any official scientific basis; it had been proposed to and accepted by the European Union on public-relations grounds, to "restore public confidence in beef." After an emergency Cabinet meeting in late September 1996, citing "the new scientific evidence" for support, John Major announced that Britain would halt its slaughter of older cattle and would not begin culling younger animals pending further evaluation. Douglas Hogg faced down the farm ministers of the EU bluntly; the predictive study, he told them, showed that "BSE will in any event die out in 2000 or 2001" and that no slaughter or cull "would make a substantial difference to the rate of decline."

In the meantime, the beleaguered British government, stung by the popular and EU response to its March vCJD rev-

---

*These, of course, would not reveal themselves until four or five years later—1998 or 1999.

elations, had locked up its data. Several researchers told me they had been warned of prosecution under the Official Secrets Act if they revealed information publicly. The predictive study that Major had cited to justify halting the slaughter program had used previously confidential MAFF statistics to arrive at its optimistic results. But word filtered through the scientific community in autumn 1996 that the maternal-transmission study on which those results were partly based was botched and that *Nature* had rejected it for publication. The calves used in the study—some the offspring of cows with BSE, some from cows not visibly afflicted—had been fed diets that might have been contaminated with BSE or scrapie. The supposedly healthy "control" cows and calves came from the same farms and herds as the infected animals and were therefore not certainly free of the disease. The study, which cost £6 million, therefore neither proved nor disproved maternal transmission and certainly could not establish a rate. Nor was horizontal transmission of BSE excluded. The predictive study played down infection from animal to animal, asserting that "no evidence exists to support the notion that the BSE agent can be transmitted horizontally either through close contact between susceptible and infected animals or through contaminated pasture," but admitted that "this route cannot be eliminated, given evidence for the clustering of cases in individual herds." There was abundant evidence suggesting maternal and horizontal transmission in the epidemic statistics: according to MAFF, 28,402 animals born after the ruminant-feed ban had sickened with BSE. Richard Lacey and a colleague had already reported evidence of paternal transmission via semen as well.

The predictive study had estimated that more than seven hundred thousand BSE-infected cattle had entered the human food supply in Britain up to the end of 1995. That was more than two percent of the approximately three million cattle the British slaughtered annually: one in fifty animals. How many people might have been infected as a result? No one knew, and estimates varied wildly. Just as MAFF had restricted access to its animal studies and statistics, so did Robert Will of the CJD Surveillance Unit restrict information on his ongoing investigation of new suspect cases of vCJD. Another British case was diagnosed retroactively after Dorrell's March announcement. But when French scientists reported two additional, highly suspect French cases in spring 1996, Will refused to reciprocate with information on further British cases he might be investigating. Someone leaked the information: the London *Sunday Times* reported in June that beyond the eleven previously confirmed in Britain and one in France, the CJD Surveillance Unit was following five more suspect cases in the summer of 1996.

Carleton Gajdusek called me in mid-July, sounding apocalyptic. "They don't have the least idea what caused the human cases," he told me. "It's kuru and nothing but kuru, and any species could be carrying it—dairy cows, beef cattle, pigs, chickens. They need to assess the risk and deal with it realistically. All the pigs in England fed on this meat-and-bone meal. The disease hasn't turned up in pigs only because you don't keep pigs alive for seven or eight years; they're killed after two or three years at most. When we kept pigs we'd inoculated in our laboratory for eight years, they came down with scrapie. Probably all the pigs in England are infected. And that means not only pork. It means your pigskin wallet. It means catgut surgical suture, because that's made of pig tissue. All the chickens fed on meat-and-bone meal; they're probably infected. You put that stuff in a chicken and it goes right through. A vegetarian could get it from chickenshit that they put on the vegetables. It could be in the tallow, in butter—how the hell am I supposed to measure infectivity in butter? No one on earth knows how to do that. These people who've come down with CJD have given blood. It's undoubtedly in the blood supply. The answer in that case is, stop giving anyone blood who doesn't really need it. Bob Will and those people don't know anything about where it came from. But I'll tell you this. If it turns up in one kid under fifteen, it's kuru. And by the way, it could be in the milk. That hasn't been excluded either."

If the new French and British cases were confirmed, the total would increase to eighteen cases in as many months. Worse, a minimum incubation period of ten years, a reasonable estimate, would put the origin of those cases back at the beginning of the BSE epidemic, when the number of infected animals entering the human food supply was small—implying that many more deaths might follow from the increasing human exposure to infected beef in the later 1980s. In October 1996, variant CJD was diagnosed in a thirty-three-year-old British woman, bringing the number of confirmed cases to thirteen. A fourteenth victim was added to the list a few days later when the Hôpital Neurologique in Lyons reported determining that one of the two suspect French cases, a fifty-two-year-old woman from Savoie who had died in August 1995, showed variant CJD pathology. To finish out a somber month, Dr. John Collinge, a neurologist at St. Mary's Hospital in London, and colleagues including James Ironside reported in *Nature* that the molecular signature of BSE in cattle matched the molecular signature of variant CJD: "'New variant' Creutzfeldt-Jakob disease," the researchers reported, ". . . has strain characteristics distinct from other types of CJD . . . which resemble those of BSE transmitted to mice, domestic cat and macaque, consistent with BSE being the source of this new disease."

At least one of the suspect cases in Robert Will's restricted portfolio was part of a cluster of four patients in East Kent, where veterinarian Colin Whitaker had found the first reported BSE-infected cow in 1985. Two of the four patients in the cluster had already died. Dr. Mathi Chandrakumar of the East Kent Health Authority assessed the meaning of the cluster pessimistically. "The annual incidence of CJD cases worldwide," he told British television, "is approximately one case in a million population. So in the approximately quarter-of-a-million population including the Ashford and Canterbury area, I would expect one case every four years or so. And we have had four cases in one year. This may be one of the many clusters we may be seeing, of many more clusters to come." SEAC's Jeffrey Almond admitted that "the worst-case scenario is possible. We desperately hope the worst case will not come about. But we have to acknowledge that it might."

When I interviewed Richard Lacey in the sunny conservatory of his house at the end of a lane in the country outside Leeds, I asked him what the worst-case scenario might be. Younger cases with a shorter incubation period had shown up first, he pointed out, just as they had among the Fore. That parallel had led him to look to kuru for a model. With kuru, the source of infection had been cut off abruptly when the Fore had given up cannibalism. As a result, cases had declined even as incubation periods had lengthened; by 1996, fewer than six Fore a year still sickened with kuru, after an incubation period above forty years.

With BSE, Lacey pointed out, there was no certainty that the source of infection had been cut off; indeed, there was evidence that animals were still becoming infected with BSE and every reason to suspect that animals incubating BSE were still entering the human food supply. The consequences, he feared, could be dire for the British as they had been dire for the Fore. "If it seems that the incubation-period average for CJD in humans begins to be about twenty-five years, maybe thirty years," he told me grimly, "then the peak human epidemic will come around the year 2015. If the current numbers of variant CJD cases increase by fifty percent per year compound, as they well might, that would take it to about two hundred thousand cases a year by then." *Human* cases, that is: 200,000 deaths *per year.*

• • •

In 1989, to prevent the spread of BSE to the United States, the U.S. Department of Agriculture prohibited the importation of British cattle and zoo ruminants. The USDA found 499 animals on hand at that time that had been imported before the ban. Those were quarantined. Two American rendering associations asked their members in December 1989 voluntarily to discontinue processing fallen and sick U.S. sheep. In 1990, the USDA began active surveillance for BSE, examining the brains of several hundred downer cattle each year for signs of spongiform encephalopathy. In 1992, the U.S. Food and Drug Administration (FDA) surveyed rendering practices and found, an FDA veterinarian writes, that "the rendering industry voluntary ban did not appear to be fully implemented since 6 of the 11 renderers processing adult sheep with heads are selling rendered protein by-products to cattle feed producers." The FDA's concern with sheep processing reflected an assumption that BSE derives from scrapie—the unsupported assumption that also misdirected British efforts. In 1993, the FDA sent letters to drug, dietary-supplement, animal-product and other manufacturers urging them to use cattle materials only from non-BSE countries in their products. In 1994, still pursuing scrapie, the USDA proposed banning sheep offal from animal feed, but U.S. agribusiness contested the ban on the grounds that such processing would be too expensive and it was not imposed.

With Stephen Dorrell's announcement in March 1996 of a possible link between BSE and new-variant CJD, the USDA ordered the 116 imported British animals still alive in the U.S. destroyed. The FDA announced in May that it would prepare regulations to prevent BSE in U.S. cattle. Livestock industry and veterinary groups agreed to institute a voluntary ban on feeding ruminant protein to ruminants, but no inspection system verifies compliance. The U.S. Centers for

Disease Control began a pilot surveillance for vCJD in five states where emerging-infectious-disease programs were already operating. The USDA decided to examine more downer-cattle brains; by late 1996 the total since the program began in 1990 had reached 3,200. None showed a BSE pattern of lesions, but they represented only a marginal sample of the approximately one hundred thousand downer cattle processed annually in the U.S. A meeting the USDA sponsored in the wake of the Dorrell announcement of seventy U.S. animal and public health experts nevertheless declared itself satisfied that existing safeguards were adequate. The group expressed satisfaction on two grounds: because BSE had never been identified in the U.S. and because the annual incidence of CJD had held steady at about one case per million population since 1979, with deaths among people under thirty years of age extremely rare.

One expert demurred from this optimistic assessment: Dr. Richard F. Marsh of the University of Wisconsin in Madison. "The fact that we have no reported cases of bovine spongiform encephalopathy in this country," Marsh told *The New York Times,* "is a false sense of security because it is not based on sufficient testing. If the United States ever gets mad cow disease, we will have the same kind of reaction that occurred in Britain." Marsh, who chairs the department of animal health and biomedical sciences at Madison, had reason to question the U.S. government's complacency. He was an expert on transmissible mink encephalopathy (TME), and he had investigated an outbreak in 1985 that he believed could only be attributed to feeding mink the meat of downer cattle.

"In April of 1985," Marsh writes, "a mink rancher in Stetsonville, Wisconsin, reported that many of his mink were 'acting funny,' and some had died. At this time, we visited the farm and found that approximately 10 percent of all adult mink were showing typical signs of TME: insidious onset characterized by subtle behavioral changes, loss of normal habits of cleanliness, deposition of droppings throughout the pen rather than in a single area, hyperexcitability, difficulty in chewing and swallowing and tails arched over their backs like squirrels." Progressive deterioration followed: loss of coordination, "long periods of somnolence in which the affected mink would stand motionless with its head in the corner of the cage," increasing debilitation leading to death. Sixty percent of the total Stetsonville breeding herd of 7,300 adult animals sickened and died. Neuropathological examination confirmed TME.

Because previous outbreaks of TME had been traced to contaminated feed, Marsh had questioned the rancher closely. "He used commercial sources of fish, poultry and cereal," the Wisconsin scientist reports. "Most of the fresh-meat portion of the ration came from fallen and sick dairy cattle which were picked up within a 50 mile radius of the mink ranch and returned for processing (butchering, grinding and freezing); a few horses had been used. *Sheep products were never fed to the mink* and there were no feed supplements of meat-and-bone meal."

Marsh had never been able to infect mink with sheep scrapie by feeding. Even direct inoculation of scrapie into the brain produced TME only occasionally in mink, with nothing like the virulence of the ranch outbreaks, and the incubation period was longer. Marsh wondered if he could passage TME to cattle and from cattle back to mink. To find out, he injected Stetsonville mink brain into the brains of two six-week-old Holstein bull calves. Each bull developed a fatal spongiform encephalopathy a year and a half after inoculation. That wasn't particularly remarkable. But Marsh then successfully passaged the disease from the bulls back into mink, by direct brain inoculation but also *by feeding the mink whole brain tissue*. And the disease agent that had been passaged through the cattle was no less deadly than TME agent passaged directly from mink to mink. "This suggests,"

Marsh reported ominously, "that there are no species-barrier effects between mink and cattle. . . ." Marsh concluded that if mink on the Stetsonville ranch were exposed to TME by being fed downer cattle, "there must be an unrecognized scrapie-like disease of cattle in the United States."

Marsh, Bill Hadlow and several other researchers decided to study the interactions of spongiform disease in mink and cattle further after Marsh's experiments with bull calves. In 1990, Hadlow retrieved from his laboratory freezer two batches of mink brain homogenate that had been stored since the 1963 TME outbreaks in Blackfoot, Idaho, and Hayward, Wisconsin. These twenty-seven-year-old inoculants produced a spongiform encephalopathy in steers that looked exactly like the new disease the Stetsonville material had produced—suggesting that the same strain was involved in all three even though they had been collected at sites distant in space and time. The damage was more extensive than the damage the British were seeing in BSE. "It affected other areas that BSE doesn't," Hadlow told me, "high up in the brain, way up here in the frontal part." When Hadlow showed cross sections of these steer brains to a British expert on BSE, the expert thought the damage looked much more like the damage produced by experimental BSE—BSE transmitted by brain inoculation, just as these had been, rather than by feeding. More ominously, the animals showed only slight signs of illness during most of the course of the infection. "Only in the advanced stages of disease," the scientists concluded, "would [the illness] have been noticed readily under field conditions." That meant the disease could easily be missed in cattle until they went down.

Then Hadlow, Marsh and their colleagues passaged BSE from British sources into mink, both by direct brain inoculation and by feeding. Scrapie hadn't taken in mink by feeding. BSE did. The resulting encephalopathy resembled the American mink disease, but it wasn't identical—suggesting that the British strain of BSE was different from the American.

Are U.S. cattle a natural reservoir for an American strain of BSE? A better question might be: if Creutzfeldt-Jakob disease occurs sporadically in one in a million humans throughout the world, shouldn't there be correspondingly rare, sporadic spongiform diseases of other animals as well? One intriguing hint is a recent estimate based on the history of mink encephalopathy outbreaks in Wisconsin—four of the total of five that have occurred in the U.S. since 1947. If every Wisconsin outbreak of TME was caused by feeding downer cattle, then the prevalence of spongiform encephalopathy in those cattle was 1:27,500, which extrapolates to one case annually per 975,000 adult cattle in Wisconsin—a ratio strikingly like the ratio of 1:1,000,000 for sporadic CJD.

A French veterinarian described what he called "a case of scrapie in an ox" as early as 1883. The first U.S. outbreak of mink encephalopathy preceded the arrival of scrapie in the U.S., another indication that sheep were probably not the source. In 1980, Colorado State University pathologists reported a "spontaneously occurring form of spongiform encephalopathy" that killed fifty-three captive mule deer and one black-tailed deer between 1967 and 1979. The disease transmitted experimentally to other mule deer. A similar disease turned up in captive elk pastured next to the mule deer, suggesting horizontal transmission. Since 1992, this "chronic wasting disease" has been identified in Wyoming deer and South Dakota elk. "The Department of Agriculture hates to hear me say this," Joe Gibbs told me, "but spongiform encephalopathy in mule deer and elk isn't surprising to me. I've seen a letter from an eighteenth-century veterinarian to a physician describing a scrapie-like disease in deer in an English deer park. So my concept is that these diseases occur spontaneously in nature. It's a rare occurrence. It would normally be overlooked, because nobody cares about an ataxic rabbit [a rabbit with loss of coordination, that is] or an ataxic squirrel or an ataxic deer. If they do, they immediately think of rabies, and if they don't find evidence of rabies, they don't care what the rest of the brain looks like." Richard Marsh came to the same conclusion about BSE in cattle in 1988: "Because this new bovine disease in England is characterized by behavioral changes, hyperexcitability and aggressiveness, it is very likely it would be confused with rabies in the United States and not be diagnosed." By 1996, Marsh had concluded, "If spontaneous cases of prion diseases can occur in humans, they likely also occur in animals. Normally not naturally transmitted, these spontaneous incidents can still pose a danger by the unnatural act of cannibalism as seen in kuru in humans, or by the intervention of man and the feeding of animal protein to ruminants."

The conditions exist in the United States for an epidemic of a native strain of BSE. There are about a hundred million cattle in the U.S., compared to fewer than ten million in Britain. U.S. renderers pick up downer cattle, some one hundred thousand per year, and process them into meat-and-bone meal. The U.S. produces about 3.6 million tons of this animal protein annually (incorporating not only downer cattle but also slaughterhouse wastes), almost ten times as much as the British produce. At the beginning of 1997, the U.S. livestock industry fed about thirteen percent of this tonnage to cattle, about the same as the British do. Meat-and-bone meal had previously been fed to adult cattle more than to calves—to beef cattle to fatten them for market and to dairy cows to support milk production—but that practice was changing in the U.S. Beef cattle are typically slaughtered at two years of age or younger, when they might be incubating spongiform disease but would be unlikely to show symptoms. Feeding animal protein supplements to calves increases their exposure over time to any disease agents that might be lurking in their feed.

Cattle protein is recycled cannibalistically in the United States, as it is in Britain. Pig and poultry protein is also recycled cannibalistically through those animals. Pigs are susceptible to spongiform disease, including BSE. Chickens may be: their brains produce a chicken form of PrP, and British researchers who inoculated chickens experimentally with BSE reported in 1991 finding cerebellar holes in one chicken's brain but no SAF. More troubling is the possibility Gajdusek cites: that meat-and-bone meal, passing through the chicken gut with any BSE infectivity unchanged, might contaminate organic vegetables for which it is commonly used as fertilizer—putting even vegetarians at risk.

Will BSE come to America? The answer seems to be: it's already here, in native form, a low-level infection that industrial cannibalism could amplify to epidemic scale. We still feed meat-and-bone meal to cattle. And an estimated seventy-seven million Americans eat beef every day.

Richard Lacey thinks his countrymen should slaughter their entire cattle herd, all ten million animals, and give up eating beef. If they did that, they'd probably have to slaughter their sheep as well. Neither prospect is likely, unless a new plague of vCJD ensues.

A story in the London *Observer* published only days after the announcement in Parliament in March 1996 of human transmission of BSE imagined just such a worst-case scenario, twenty years down the road:

> It is 20 March, 2016. . . . Now, Britain's National Euthanasia Clinics churn on overtime, struggling to help 500 people a week to a dignified death before brain disease robs them of reason and self-control.
>
> This nation, whose leaders spent a decade in denial, is now quarantined, the world having long since shunned contact with a population in which half a million people a year succumb to Creutzfeldt-Jakob disease, a fatal neurological illness spread in the late twentieth century

> through the eating of infected beef products. The Channel Tunnel is blocked with five miles of French concrete. The health service is crippled; blood transfusions are impossible because undetectable prions . . . infect most donors, and the strain of caring for more than two million CJD victims has overwhelmed support staff. The fabric of the nation is being torn apart. . . .

Britain may well face such a grim prospect, though how many people will become infected remains to be seen. The British government, by making the wrong public-health choices, has conducted a frightening natural experiment, allowing a lethal disease agent to spread through the human food supply, exposing the entire British population. There is every reason to believe exposure is continuing, from infected beef and possibly from infected lamb and mutton as well.

At some lower level of risk, the population of Europe has also been exposed to BSE. The lack of vCJD deaths outside France doesn't mean that the rest of the continent is free of human infection; the French deaths indicate merely that exposure began there earlier. Nor do the limited number of vCJD deaths identified so far offer any basis for assessing the future extent of vCJD infection. That depends on the virulence of the disease agent. The ease with which BSE crosses species barriers and the fact that it is easily transmissible orally in small doses suggest that it may be an exceptionally virulent strain.

The United States is probably not currently at risk from the *British* strain of BSE. That's the good news. But livestock feeding practices put the U.S. population at risk from a *native* strain of BSE, which could spread quickly through the livestock population in contaminated feed and infect humans just as British BSE has done. Americans would then face the same prospect the British are facing: of a plague of fatal brain disease that might kill hundreds of thousands of men, women and especially children every year.

What about the rest of the world? If TSEs occur sporadically in animal species, as CJD does in humans, then no population anywhere in the world that eats meat is entirely free of risk.

Occurrences of most epidemic diseases continue to decline from decade to decade in the advanced industrial nations. In the U.S., for example, with the exception of AIDS, most diseases that physicians are required to report to the Centers for Disease Control—equine encephalitis, brucellosis, diphtheria, gonorrhea, hepatitis B, measles, mumps, polio, rubella and syphilis, among others—have declined in the past four decades or disappeared. The mortality rate in New York City at the beginning of the twentieth century was comparable to the mortality rate today in Bangladesh, but few people in the U.S. or in Europe any longer have personal knowledge of the ancient killer diseases of humankind—bubonic plague, smallpox, cholera, malaria. Good nutrition, improved sanitation and hygiene and new vaccines limit the risk of these infections.

But the British BSE disaster was not some throwback to antique conditions or mere bad luck. It followed in part from the pernicious, pervasive and deeply corrupt antigovernment fanaticism that has taken hold in Britain and the U.S. in the last two decades, the same benighted mentality that decrees that illegal immigrants should be denied medical treatment when they are ill. There's nothing inherently wrong with industrializing agriculture—with raising chickens, pigs and other food animals under controlled conditions—*if* such conditions are carefully arranged to limit disease and protect the public health. Human beings, after all, have chosen to live under controlled conditions. We thrive with a roof over our heads and a healthy food supply; so do farm animals raised in confinement, which contract fewer diseases and survive to maturity in greater numbers than their "free-range" counterparts. But eggs and chicken meat contaminated with salmonella, hamburger and fresh cider poisoned with deadly *E. coli* and beef infected with BSE all evidence a failure of government inspection, supervision and policy.

Until 1995, U.S. meat inspection had hardly changed since the reforms that followed Upton Sinclair's 1906 fictional exposé *The Jungle*. Even with the passage in 1995 of a modernized meat-inspection system that for the first time included laboratory testing, the radical Republican Congress under Newt Gingrich and Bob Dole refused to exempt the measure from the so-called regulatory relief provisions of its Contract with America, delaying and limiting its effect. The Congress was demonstrably more interested in meat-industry profit margins than in consumer protection.

The British disaster makes it grimly evident that the high-tech cannibalism of recycling animal protein risks spreading lethal disease. The practice continues in the United States for the same reason that it continues in Britain even as the death count rises: because the meat industry and its allies in government assess the risk differently from the scientists and physicians who know most about the transmissible spongiform encephalopathies.

The U.S. still has a chance to avoid a BSE epidemic. The FDA took an encouraging step in that direction at the beginning of 1997 when its director, Dr. David Kessler, announced that the agency proposed to ban the feeding of ruminant protein to ruminants. The action opened a forty-five-day period of public comment, after which the ban was expected to take effect. The U.S. rendering industry immediately complained that the agency was overreacting. Veterinarian Don Franco, a spokesman for the National Renderers Association, told *The Wall Street Journal*, "The science just isn't clear enough. There are lots of theories, but there hasn't been any specific proof on how the disease is transmitted."

The *Journal* reported an industry estimate that a ruminant-protein ban would cost $1.6 billion in lost sales and for alternative disposal. Disposal of banned wastes could be a predicament as difficult for the U.S. as it has been for the British. Dennis Longmire, head of the thirty-two-plant U.S. rendering firm of Darling International, characterizes the problem grimly: "You're talking about incinerating or land-filling millions and millions of tons of perishable animal tissue a year." The National Cattlemen's Beef Association, on the other hand, welcomed the move, claiming that beef and dairy producers had already reduced feeding meat-and-bone meal to cattle.

But there were disturbing compromises built into the FDA's proposed ban. It would prohibit the use of tissue from cattle, sheep and goats in feed for those ruminants but would permit the continued feeding of ruminant blood, milk and gelatin. Ruminant tissue would continue to be processed into feed for chickens, pigs and pets despite the known susceptibility of pigs and cats to spongiform encephalopathy and the possible passage of the disease agent through chickens into their manure. Nor do the ubiquitous contamination of U.S. poultry and eggs with salmonella and the continuing outbreaks of human *E. coli* infection from contaminated meat inspire confidence in government inspection. From the rendering industry's perspective, the ban turns a valuable asset into an expensive liability. That reversal in Britain led to widespread cheating. Without rigorous enforcement, U.S. consumers have reason to fear the same result.

The U.S. has ample supplies of soybean protein, so its livestock industry could survive even a complete ban on feeding animal wastes to animals. Such a complete ban is the only certain way to prevent the passage of fatal spongiform encephalopathy among species—including, pointedly, human beings. The industry's profit margin would narrow, since meat-and-bone meal is cheaper than soybean meal. But such an embargo is not even under discussion in Britain, much less in the United States. It would mean disposing of millions of tons of animal wastes left over from meat processing—a nearly intractable problem. Alternatively, we could follow Richard Lacey's advice and give up eating beef.

The next major breakthrough in medicine is expected to be xenotransplantation: grafting animal tissues and organs into humans. Some twenty thousand transplant operations are performed each year in the United States, but forty thousand Americans beyond that need heart, liver, kidney and cornea replacements, and ten thousand of those die before they reach an operating room. In other countries, transplant organs are rationed and many patients who could benefit from them suffer or die as a result. Xenotransplants could improve and extend all these lives.

Pioneer xenotransplantation has already begun: in 1984 in the U.S., a baboon heart kept Baby Fae alive for twenty days; a baboon liver was transplanted in 1994; San Francisco AIDS patient Jeff Getty received a baboon-marrow graft in 1995 to shore up his immune system. Advanced biotechnology that may make xenotransplantation routine is under development in the United States and in Britain. Lines of transgenic pigs are being bred for use initially for heart transplants. Pig blood types are more like human types than those of other animals, but a strong immune response known as hyperacute rejection normally destroys pig tissue grafted into primates in a matter of hours.

I investigated Imutran, a company based in Cambridge, England, that leads the world in xenotransplantation technology, and learned that it has cloned human genes that defeat hyperacute rejection and inserted them into pig embryos. Imutran has bred hundreds of pigs carrying these human genes. Rejection of transgenic pig hearts still has to be controlled with drugs, just as rejection of transplanted human hearts has to be controlled with drugs. In 1995, Imutran demonstrated that even without such immunosuppressive drugs, monkeys implanted with its transgenic pig hearts survived for five days—well past the time when hyperacute rejection would have destroyed an ordinary pig-heart implant. Implanted monkeys treated with immunosuppressive drugs survived up to sixty days. That achievement led Dr. David White, Imutran's cofounder and medical director, to predict routine pig-heart transplants in humans before the turn of the century. "The big debate now," White told the media, "is, do we currently have the skills to keep the hearts functioning in people for a long time; and the only way to answer that question is to put them into people and find out."

I interviewed White at Imutran's headquarters in Cambridge in April 1996. He was enthusiastic about his work. "Right from the beginning," he explained, "our approach was to ask how can we genetically engineer the pig, not how can we treat the patient. From there, we realized that a possible approach would be to put these human regulators into a pig. And the smartest thing I ever did was to take out a patent on the process. Because that's what pays all the bills." Although I didn't know it at the time, White had just sold Imutran to Sandoz Pharma, Ltd., a major drug company.

"I will put my career on the line," he told me, "and say that hyperacute rejection as we know it is dead, gone, finished. You take an organ from one of our pigs and transplant it into a primate and it will go for days without any treatment at all, routinely. We've done kidneys, islets [i.e., pancreatic tissue which secretes insulin, to correct diabetes], hearts—I don't even know the number now, sixty or seventy. Now all we have to do is immunosuppress the monkey in order to achieve long-term survival. We did our first baboon transplant a couple of weeks ago, and on the same day that we successfully transplanted a baboon with a pig heart, one of our patients died waiting for a human heart."

I came to the point of my visit: "Are you concerned with BSE?"

"Fortunately," White countered, "pigs don't get BSE."

"I think there's evidence they do."

"If you take contaminated brain from a mad cow and inject the neural tissue directly into the brain of the pig it will get spongiform encephalopathy. But they've been feeding infected brain to pigs for five years now and none of the pigs has the disease."

That was true.

"You have to appreciate that BSE is not an infection. It's a very curious toxicity really."

I told Dr. White I'd looked into it.

"Well," he responded, "then perhaps you can tell me how the hell the bloody thing works. I don't understand it."

I tried to explain abnormal protein crystallization.

He listened. "Yes, that could work," he said finally.

"Your pigs are isolated and presumably not fed meat-and-bone meal," I prompted him.

"Oh no," he confirmed. "Disease transmission is an area of considerable concern." He left his desk and returned with a proprietary study as thick as a telephone book. "We put together a group of the world's leading experts on pig disease and on the diseases that transplant patients get." He opened the book. "I'll just read you some of the headings. 'Microorganisms Known to Be Pathogenic.' 'Microorganisms Pathogenic in Humans.' 'Microorganisms Known to Be Pathogenic in Pigs but Not Pathogenic in Humans.' 'Microorganisms Not Known to Be Pathogenic but Similar to Microorganisms Pathogenic.' And so on. Porcine RNA viruses, porcine DNA viruses, exotic porcine RNA viruses, exotic porcine DNA viruses, a special section on the human measles viruses. Porcine bacteria—the gram negatives, the gram positives—and it goes on and on. A basic risk assessment of them all. A list of pathogens of most concern." He closed the book. "So when you've done all that, you're left with one problem, which is the retroviruses. We're currently doing research on our primates to answer the question, will these pig retroviruses jump across the species barrier and recombine with human retrovirus? We haven't finished, but we think the probability is extremely remote."

The pigs probably won't go to the hospital, White continued. The patient will come somewhere near the pigs. "That is," he explained, "you will have a few dedicated specialist centers which do xenotransplantation. Those centers will have a sterile pig-production unit nearby. The patients will come there. It is ludicrous that you have to wait for fit, healthy people to die so that you can treat sick people. With a pig, you can come in and the physician will say, 'I think you're going to need a heart transplant.' You wouldn't be at the end of the road. Maybe three months, maybe six months away. And we would modify *your* immune system so that you won't reject pigs."

It occurred to me that we might be talking about more than hearts. "Are you planning to transplant other organs and tissues from the pig?" I asked.

"The heart, the lungs—all those former smokers, the market is huge—the kidney. Possibly the intestine. The substantia nigra is an area of great interest."

I said: "What?"

"Bit of the brain," White said. "For the treatment of Parkinson's disease."

I knew what the substantia nigra was. I just couldn't believe that a brilliant and innovative physician-businessman who had admitted he didn't understand what causes spongiform encephalopathy (who does?) was planning to implant pig brain directly into the brains of humans.

In July 1996, the Committee on Xenograft Transplantation of the U.S. Institute of Medicine, part of the National Academy of Sciences, endorsed xenografting on the grounds that the potential benefits outweigh the risks. "When the science base for specific types of xenotransplants is judged sufficient," the committee concluded, "and the appropriate safeguards are in place, well-chosen human xenotransplantation trials using animal cells, tissues and organs would be justified and should proceed." The committee cited "ample evidence," however, that infectious agents could be transmitted from animals to humans, which indicated a danger "unequivocally greater than zero" that xenotransplantation could transfer new and deadly viruses across the species barrier. And it specifically mentioned transmissible spongiform encephalopathy.

If I expected to die of heart disease in a matter of months, or needed a new set of lungs, I would certainly pursue a pig transplant. I might be leery of a cornea or a bit of the brain. Until the cause of TSE is identified, xenotransplantation will continue to carry some degree of fatal iatrogenic risk.

Even the origin of *sporadic* CJD continues to be debated. A British researcher, Dr. Richard Kimberlin, points out that the age-specific incidence of sporadic CJD:

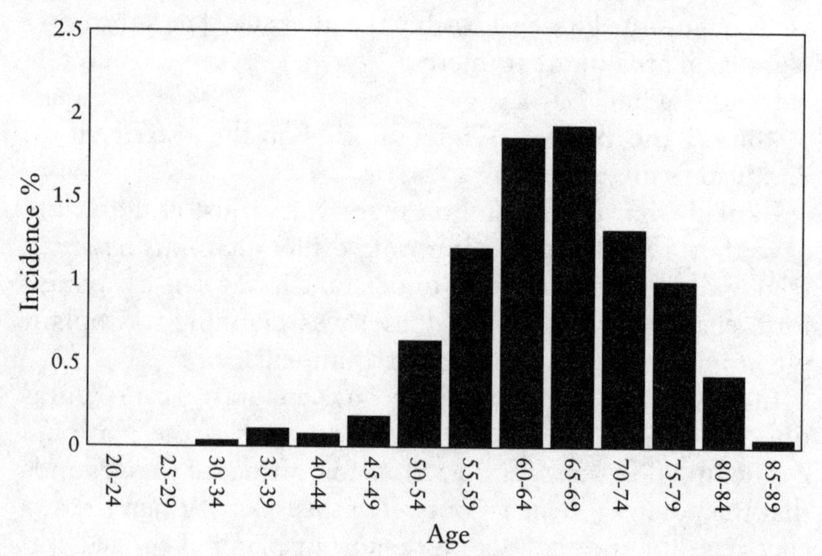

is similar to the age-specific incidence of BSE:

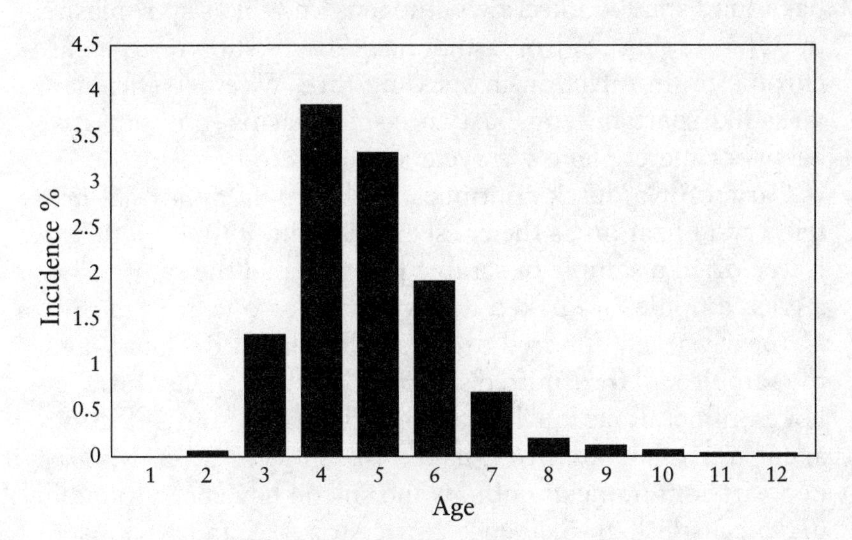

If either disease originated in a mutation or some other random event, incidence should increase steadily with age. The bell shape of the BSE age-incidence curve implies a different scenario: infection at an early age followed by a period of incubation. The length of the incubation period depends on an animal's individual resistance—shorter in some animals, longer in others. The peak of the curve marks the median incubation period—the halfway point between the shortest and the longest. Kimberlin argues that the bell curve of sporadic CJD indicates that this seemingly random human disease is probably *also caused by infection:* "The shape of the age-specific incidence curve . . . implies that infection with [a] common strain [of CJD] occurs in childhood or adolescence, and that the median incubation period is 40 to 50 years." German researcher Dr. Heino Diringer similarly defends an infectious cause: "It seems more than likely that . . . the sporadic cases of CJD always originate from direct or indirect transmission from animals to man." In 1996, Diringer reported finding small virus-like particles in scrapie hamster brain. An American neuropathologist, Dr. Frank O. Bastian, has found small corkscrew pathogens known as spiroplasma in CJD brains. Spiroplasma have been shown to cause chronic brain infection in suckling rats. Whether Diringer's virus-like particles or Bastian's spiroplasma are causative agents or merely fellow-travelers remains to be seen.

Carleton Gajdusek continues to champion abnormal protein crystallization as the cause of TSE. In 1990, Paul Brown freeze-dried a sample of scrapie brain, sealed the sample into a glass ampule and baked it in an oven for one hour at 360° C. (nearly seven hundred degrees Fahrenheit). Reconstituted, the sample still transmitted scrapie to a hamster. Gajdusek invokes minerals as nucleants to explain how an infectious agent might survive being baked to ash at 360° C. Mineral deposits containing aluminum and silicon have been found in nerve cells in high-incidence clusters of ALS and Parkinsonism-with-dementia cases on Guam and other Western Pacific islands. Aluminum silicate deposits form the cores of Alzheimer's amyloid plaques. They may be crystals of a common form of clay, montmorillonite. Alzheimer's researchers have proposed, Gajdusek writes, "that they are the initiating elements of the amyloid deposition." It was this finding of aluminum silicate cores in Alzheimer's plaques that led to a panic a decade ago among users of aluminum cookware, although no connection between cooking in aluminum and Alzheimer's disease has ever been confirmed.

Aluminum and silica are common elements in the earth's crust. Clay is common throughout the world. How might common mineral crystals "infect" one in a million human beings with sporadic CJD? Gajdusek described to me a slide he sometimes uses in lectures to pathologists, a micrographic cross section of a deep interior part of the brain called the hippocampus, its cells peppered with what look like black basketballs. When the pathologists give up guessing, Gajdusek tells them that the black basketballs are particles of gold dust. When they give up guessing how gold dust found its way deep into the center of the brain, he explains that he asked a dying patient to sniff it after giving the man scopolamine to dry up the lining of his nose. "Reverse axonal transport," Gajdusek names the process, meaning the gold dust is drawn into the nerve endings in the back of the nose that collect molecules of smell information out of the air, whereupon the nerve fibers—the axons—transport the particles to the hippocampus, the place where smells are deciphered.

We were sitting in the kitchen of Gajdusek's suburban house in a valley on the outskirts of Frederick, Maryland—he gave up his hilltop mansion in 1989. "Did you know you can kill someone with a handful of glass dust?" he asked me. I shook my head—I didn't know that. "The Germans studied it in the First World War as a possible war gas. Glass particles of a certain size will lyse [disintegrate] blood cells better than cobra venom. One gasp and you're dead—the lungs fill up with blood. But you can also *block* the effect with glass dust. Finer particles will bond to the sites on the cell walls and prevent the killer-sized particles from docking."

Gajdusek leaned forward and slammed both his hands flat on the table, his way of making sure he has your attention. "*All* the different theories about the nucleation process of the infectious amyloids are the same. It's like theories of ice formation—one says dust causes ice crystals to form, another says ice-crystal fragments, another says pieces of locust wing. Bullshit. They're all the same theory. These are fractal events. And there's no general theory of nucleation. My colleagues want a simple formula. It's too complex. You need a little piece just big enough to give the pattern. In the real world of nucleation, one particle of dust in a million can change the outcome. Every crystallization on earth requires nucleation—a blueprint—and that includes many bodily processes. But there's nothing at all in the medical literature about nucleation. It's a completely overlooked phenomenon."

I thought: nature's boundless depths: layers under layers down into the very center of things, and layers there too small to see, and layers below those layers until the head swims and still more layers then. If an environmental agent such as mineral dust initiates amyloid formation in TSE, then the spongiform diseases will always be with us, like fossilized remnants of Original Sin ground fine in God's mortar. Or like Camus's plague lurking in the sewers, waiting for us to nod.

Listening to seventy-four-year-old Carleton Gajdusek spin gold and glass and mineral clay into an explanation for fatal brain disease, I remembered something he asked me at our first meeting, late in 1995, before the British reported out the beginnings of what may be their new Black Death.

"You know the bone meal that people use on their roses?" Gajdusek asked me then. "It's made from downer cattle. Ground extremely fine. The instructions on the bag warn you not to open it in a closed room. Gets up your nose." The Nobel-laureate virologist who knows more than anyone else in the world about transmissible spongiform encephalopathy looked at me meaningfully. "Do you use bone meal on your roses?"

I told him I did.

He nodded. "I wouldn't if I were you."

From the London *Daily Telegraph*, April 4, 1996:

> Gardeners have been reminded by the Royal Horticultural Society to wear gloves and a dust-excluding mask to avoid any risk of BSE when applying a spring dressing of blood and bonemeal to roses and shrubs.
>
> Demand for beef is recovering steadily. At London's Smithfield wholesale market, the trade in better quality cuts of British beef has recovered from zero a week ago to just over half the normal figure.

Glade

June 1995–January 1997

AFTERWORD

We May Have to Face an Epidemic

One year after the publication of *Deadly Feasts* in hardcover, most of the news is bad. Whether Great Britain and Europe will suffer a major human epidemic remains to be seen. Twenty-four cases of new-variant CJD (vCJD) in young people in Great Britain and France have now been identified (twenty-two officially), and nothing that researchers have learned about vCJD and BSE encourages optimism. Since projecting the size of an epidemic requires tracking the increase in cases long enough to measure their trend, it will be A.D. 2000 or later before statisticians can make reliable estimates. The only historical model for a food-borne epidemic of transmissible spongiform encephalopathy in humans is the New Guinea kuru experience, and at the height of that epidemic the Fore were losing one percent of their population annually. One percent of the British population would be about five hundred thousand deaths per year. We may hope that BSE is not so virulent as kuru.

The United States and Canadian governments have taken steps to limit risk, but their actions represent utilitarian compromises between public health and rendering-industry profit margins, reducing the danger of a BSE epidemic in North America without fully protecting the public should transmission occur. From a government perspective, a more limited North American response is reasonable because no cattle are known to have died of BSE here. From the perspective of those of us who live here, anything less than the most effective possible barrier against the insidious spread of a stealthy, untreatable, incurable and invariably lethal disease is gambling with our lives. Would we be comfortable knowing that only partial protection had been installed against HIV contaminating our commercial blood supply, if the virus were killing people elsewhere but had not yet found its way to where we live? That's how much protection the American governments have mandated against animal TSEs.

The most disturbing new case of vCJD is that of a twenty-four-year-old Englishwoman, Clare Tomkins, of Tonbridge, Kent, reported in August 1997. Tomkins was a vegetarian who had not eaten meat in eleven years, a hiatus that moves the probable time of her infection back to 1986 or earlier, when very few British cattle had begun showing symptoms of BSE, although many—up to 54,000 head, according to an Oxford University Department of Zoology estimate—must have been incubating the disease. Such early transmission could mean the BSE agent is particularly virulent, so that even an extremely low dose infects. Alternatively, it could mean that animals are infectious early in the incubation period of the disease, long before they begin to show symptoms. Or both possibilities could be true. Either, or both, undermine the preventive measures so far taken.

The British have banned the feeding of any mammalian protein to food animals, for example, but the European Commission, in a study conducted in 1996, found that animal feed from British feed mills was contaminated with about five percent banned material, which had lodged in the crevices of the milling machines. If an extremely low dose is potentially lethal, how large a dose is five percent? Similarly, the British government has been slaughtering older cattle from infected herds on the unsupported assumption that the 1989 ban on feeding ruminant protein to ruminants stopped transmission, but in fact cases of BSE have continued to emerge in British herds not previously infected. Which suggests that eaters of British (and Continental) beef are still being exposed to BSE.

Despite these troubling uncertainties, two successive British governments—those of Margaret Thatcher and John Major—avoided inquiring how BSE arose in cattle or how it spread to humans. The definitive experiment would have been to feed healthy, unexposed cattle contaminated meat-and-bone meal and then follow them to see if they became infected. Such a study has not been commissioned, for reasons that remain unexplained. MAFF suppressed a 1992 study of 444 hunting hounds with neurological symptoms that found Pat Merz's scrapie-associated fibrils (SAF)—definitive evidence of TSE infection—in some animals' brains. When the cover-up was exposed, the ministry argued peevishly and falsely that public notice was unnecessary since a TSE in dogs would not affect the human food supply. To the contrary, the fact that dogs, like cats and zoo animals, are susceptible to BSE infection from contaminated food is further evidence that BSE easily jumps the species barrier. Such evidence should have been a factor in deciding how extensively to disrupt farming and commerce to protect humans from infection. MAFF's sorry record in this and similar instances may explain why it commissioned no definitive feeding experiment: among other outcomes, such an experiment could clearly assign blame.

But a major program of laboratory experiments has strongly confirmed that vCJD is the human manifestation of BSE. A team headed by Dr. Moira Bruce of the Institute for Animal Health Neuropathogenesis Unit in Edinburgh reported in *Nature* in October 1997 that mice inoculated with BSE-infected tissue from cattle, TSE-infected tissue from domestic cats, a greater kudu or a nyala, and vCJD from humans all showed similar incubation periods and similar patterns of brain damage. These signs differed in identical mice inoculated with various strains of scrapie or with sporadic CJD. "Our data," Bruce and her colleagues conclude, "provide evidence that the same agent strain is involved in both BSE and vCJD."

Supporting Bruce's experiments, John Collinge's mice finally died. When I interviewed the British research physician at St. Mary's Hospital in London in April 1996, in the wake of the first ten human deaths from vCJD, Collinge's phone was ringing off the hook, and asking after the mice's health had become the press's standard inquiry. The Cambridge-educated neurologist, who looks like a young Walter Matthau and is a member of SEAC, the government's Spongiform Encephalopathy Advisory Committee, was beginning to find the question annoying. He and his Prion Disease Group had bred the transgenic mice with their mouse PrP genes knocked out and with human PrP genes substituted, so that the animals expressed only human PrP and lacked a species barrier to human TSE agents. Collinge had then inoculated groups of these transgenic mice with different sporadic CJD, BSE and vCJD samples and settled in to await the end of incubation and the appearance of symptoms. He'd been waiting several hundred days when I saw him and the mice were still behaving normally.

In the same October 1997 issue of *Nature* as the Bruce paper, Collinge reported that his mice stopped behaving normally as they came up on five hundred days post-inoculation. Richard Marsh, the Wisconsin mink researcher (now deceased), had noticed sick minks running around with their tails arched over their backs in his investigations. Collinge found some of his mice one day persistently walking backward. The reversed mice turned out to be those infected with vCJD or BSE, but not those infected with sporadic or iatrogenic CJD or with scrapie. The brain damage in vCJD- and BSE-infected mice was similar, as were incubation periods and the length of illness once symptoms appeared. It's logical that vCJD, a human disease, should transmit to mice that make only human PrP. But the fact that BSE—mad cow disease—transmits to such mice further confirms the connection between the bovine and the human diseases. "On clinical, pathological and molecular criteria," Collinge and his colleagues write, "vCJD shows remarkable similarity in its transmission characteristics to BSE, and is quite distinct from all other forms of sporadic and acquired CJD. These data provide compelling evidence that BSE and vCJD are caused by the same prion strain. Taken together with the temporal and spatial association of vCJD with BSE but not with scrapie or other animal prion diseases, and BSE transmission studies in macaques, this strongly suggests that vCJD is caused by BSE exposure." BSE and vCJD were much more deadly to Collinge's mice than ordinary CJD.

Collinge found more subtle differences that he thought carried frightening long-term implications. Humans inherit two PrP genes, one from each parent. Since there are two versions of the PrP gene—labeled M and V—there are three types of inheritance: thirty-eight percent of humans have MM, fifty-one percent have MV and eleven percent have VV. All the human vCJD cases so far identified have been MM inheritors. The human PrP genes inserted into Collinge's mice, however, were VV. Since the VV mice nevertheless contracted BSE or vCJD, the BSE agent is evidently able to infect all three kinds of human inheritors, the two doubles and the cross. And since only one kind, MM, has so far reached the stage of showing symptoms in humans, Collinge concludes that there are likely to be three waves of vCJD epidemics. "Compared to the MMs," he told the British press, "a similar or lower proportion of VVs will probably get the disease, and probably a lower proportion again of MVs." He estimated that twenty years might pass before the second wave of infections revealed themselves, and even longer for the third. Which tends to predict a larger rather than a smaller death toll, two SEAC members implied elsewhere in *Nature:* "Much depends on the average incubation time of vCJD: the longer the time, the higher the final figure [the number of human deaths] is likely to be."

Kuru still appears among the Fore, declining now on its way to extinction through five or six cases per year, each year's victims older than the last. In June 1996, I traveled by four-wheel-drive truck on muddy, dangerous roads to the Fore region of the Papua New Guinea Eastern Highlands to interview a forty-five-year-old grandmother, Kogosa, who had begun showing kuru symptoms in May. Her village, Waisa, was far enough into the bush that when I smiled at a baby in its mother's arms it screamed in fear, having never seen a red-haired Caucasian before. Kogosa emerged from her woven-walled, thatched house in a knit cap and a brightly colored missionary dress. A muscular woman with milk-chocolate skin, she was unsteady on her feet and supported herself with a stick. We sat in the pig-redolent dust, leaning against the house wall, and talked through a translator, her adult son, who had driven down with me from Goroka.

Kogosa's mother and younger sister had already died of kuru. She was well aware that her illness was terminal, and was preparing to go into Goroka while she could still walk, to close the savings account she had accumulated over the years growing coffee. When I asked her if she had eaten the dead she looked quickly away, embarrassed, and I didn't press the inquiry. But contamination with the tissue of victims is the certain cause of kuru infection, and the slow decline in cases since the Fore abandoned cannibalism in the late 1950s is additional evidence that the practice was causative, just as industrial cannibalism of cattle by cattle caused the BSE epidemic: both epidemics were food-borne. If Kogosa participated in a cannibal feast, she had to have done so as a small child. Which meant she had silently incubated, for more than three decades, the disease that has since killed her.

Collinge's results disturbed him enough that he decided to breach the wall of silence British scientists have maintained voluntarily or under government coercion against public comment on the human implications of the BSE epidemic. "I am now coming round to the view," he told the London *Times,* "that doctors working in this field have to say what they think, even though this may give rise to anxieties which later turn out to be groundless. . . . It can no longer be denied that it is possible, even likely that we may have to face an epidemic."

It was still impossible to predict the size of the epidemic, Collinge continued—"it may only involve hundreds, but it could be Europe-wide and become a disaster of biblical proportions. We have to face the possibility of a disaster with tens of thousands of cases. We just don't know if this will happen, but what is certain is that we cannot afford to wait and see. We have to do something, right now. We have to find the answers, not only to the questions of the nature of the disease, but to find a way to develop an effective treatment." He thought a treatment might be no more than ten years away.

The chilling possibility that BSE has passed back into sheep has finally begun to be taken seriously in Great Britain. The Labour government of Tony Blair moved soon after its election in 1997 to extend BSE controls to sheep: compensation for culling diseased animals from flocks, and a "specified offals" ban on spinal cords and spleens (head having already been banned). In sheep, BSE would look like scrapie but might be infectious to humans as scrapie is not. If so, then British citizens, tragically, have been exposed to BSE seven years longer from mutton and lamb than from beef.

. . .

The same week in October 1997 that the Collinge and Bruce reports appeared, the Medical Nobel Assembly at the Karolinska Institute in Stockholm announced that it would award the 1997 Nobel Prize in Medicine to Stanley Prusiner "for his discovery of prions—a new biological principle of infection." Prusiner had campaigned relentlessly for the prize, and the announcement must have been a sweet victory, but it may prove a major embarrassment for the Nobel Assembly. All known living organisms use molecules of nucleic acid—DNA or RNA—to carry the information they need to reproduce. Prusiner has worked for years to prove that the TSE agent is an exception, unique in nature. It's an infectious protein, he says—in his coinage, which has won wide acceptance, a "prion"—that reproduces without nucleic acid. A significant number of his fellow TSE researchers disagree. In awarding him the Medicine Prize, the Karolinska group explicitly took his side in the ongoing debate, which may have profound consequences for TSE research, choking off exploration of the other possibility: that the disease agent is a virus. "There are still people who don't believe that a protein can cause these diseases," Karolinska neurologist Lars Edstrom told *The New York Times*, "but we believe it. From our point of view, there is no doubt." The vice chairman of the Nobel Assembly, Ralf Pettersson, even implied that doubt contributed to the spread of BSE to humans. "During the whole of the 1980s," he told Reuters, "the prion was very controversial. Acceptance took a while. This could have delayed moves. It was more a political decision [in Britain] about when to take action, and by then it was too late."

The religious tone of these endorsements offends neuropathologist Laura Manuelidis, a tall, elegant Sarah Lawrence graduate and Yale Medical School professor whose research turns up tantalizing hints that a virus may be involved. "No doubt?" she told me over dinner one evening in New Haven. "How could there be no doubt? Is this science?" Insouciantly picking clean a broiled fish, she counted up the major TSE labs that have serious doubts about Prusiner's prion theory and were looking for a virus: eight, about half the first-rank establishments in the field. "There must be a special strain of the TSE agent that causes egomania," she added wryly, probing the fish's head.

Readers who have come this far may find it disorienting that I'm reporting challenges to the infectious-protein theory. I raise the virus issue partly because I now believe I gave it less than a fair hearing in the body of this book and partly because the arguments I've heard in its favor since I wrote *Deadly Feasts,* especially from Laura Manuelidis, seem to me compelling. Good scientists, I've been told, let the evidence drive their theories. I'm not a scientist, but I take the point. The virus/protein question is still open, which is why the Nobel Assembly's decision to endorse Stanley Prusiner's prion version of infectious-protein theory by awarding him the most distinguished scientific prize in the world is baffling.

Bruce Chesebro, the articulate physician who directs TSE studies at the Rocky Mountain Laboratories, took the unusual step of issuing a public objection when Prusiner's Nobel was announced. Chesebro observed sharply that "no one to date knows precisely what a 'prion' really is" and specified why the virus/protein debate is important. Clumps of abnormal protein, he pointed out, are also characteristic of diseases such as Alzheimer's, diabetes and rheumatoid arthritis. Why should the prion be uniquely transmissible when other such amyloids aren't? "These unexplained issues leave open the possibility that an undiscovered virus might yet prove to be the true agent of the transmissible spongiform encephalopathies." Chesebro thought it would be "tragic to stifle the future research needed to identify the precise nature of the transmissible agent responsible for these diseases or to impede the search for drugs required to prevent or cure them."

Medicine Nobel Prizes can be awarded to as many as three people at a time; to be the sole winner is a rare distinction, which may explain why Prusiner felt comfortable telling Reuters, "This prize is ample support for what I'm saying." Since the award was supposed to be for discovering prions, Chesebro wondered why Pat Merz wasn't included. So did I, and I called Merz to commiserate when I heard the Nobel announcement. She thanked me, adding, "You aren't the first to call." Merz's discovery of scrapie-associated fibrils—Prusiner's "prion rods"—ought to have qualified her to share the prize. Significantly, Merz isn't convinced that prions carry infection. She thinks SAF probably contain nucleic acid.

Until someone identifies a nucleic-acid sequence unique to TSE infections, the best evidence for a TSE virus is the fact that TSEs come in different strains, just as flu and cold viruses come in different strains. To review: In the rest of nature, nucleic acids code strain information. Strains reveal themselves in different signs and symptoms. There are some twenty different strains of scrapie, easily distinguishable by their differing incubation periods and distinctive patterns of brain damage. Sporadic CJD is a relatively rare disease of middle age with early onset of dementia; vCJD, a different strain, stood out from that background because it affects young people and dementia develops late. On autopsy, sporadic CJD brains show widespread spongiform damage and few amyloid plaques, while vCJD brains (and BSE-infected macaque brains) show large "florid" plaques and a focus of damage in the cerebellum, the part of the brain that coordinates movement. These distinctive strain characteristics breed true; Moira Bruce, for example, identified the same BSE strain in transmissions from eight different species. The virus/protein debate, then, is a debate in part about what carries the information that makes each of these many different strains distinctive.

If prions alone carry strain information, then it ought to be possible to make synthetic PrP in a test tube and cause infection. That's been tried, and it doesn't work. Worse, PrP is different from species to species. Mouse PrP has a different amino acid sequence from cat PrP, which is different from human PrP. Bruce's transmission of the same strain of BSE from eight different species—with cight different kinds of PrP—is a serious blow to the prion camp. How could eight different kinds of PrP all code for the same strain?

Carleton Gajdusek cites the evidence that hard radiation doesn't completely inactivate TSE material as the bedrock of his conversion to infectious-protein theory; Manuelidis responds emphatically that many viruses, retroviruses in particular, are efficient at repairing radiation damage to their genome.

It's a tribute to the dazzle that Prusiner, and Gajdusek before him, have brought to the debate that the infectious-protein theory has gained the ascendancy. Prion theory predominates in medical textbooks these days; slow-virus theory is receding. That's because the lion's share of TSE research is directed toward the proteins (Prusiner's lab at UCSF has received about $40 million in funding over the years, including $4 million from the U.S. Congress "to determine the structure of prions and how they cause disease"). There's plenty of nucleic acid in the brain homogenates researchers study, but picking out the viral molecules (if they're there) from among all the cell debris is like finding a needle in a haystack. In 1982, when Prusiner proposed his prion theory, viral molecules would have been easy to overlook. If they're going to be found, they'll have to be searched out almost molecule by molecule—but hardly anyone's looking.

Laura Manuelidis reminded TSE researchers at a 1995 conference in Paris that down below the dazzle almost everything about the TSEs points to a virus, with PrP acting as a gatekeeper at some stage in its life cycle: the fact of infection, the variety of strains, their dormant persistence through long incubation periods when cells normally clear proteins quickly, and—what Manuelidis called "a hallmark of an infectious agent"—their transmission across species barriers. BSE is almost certainly transmitted orally. The stomach and gut digest protein, and no one has shown that PrP survives digestion; but viruses commonly survive the bath of hydrochloric acid and enzymatic juices to invade the bloodstream through a part of the intestine known as the distal ileum—the only part of the cattle intestine where BSE infectivity has been found.

A visiting researcher in Manuelidis's lab, Martin Brock, an associate professor of biochemistry at Eastern Kentucky University, joined us for dinner. It had occurred to Brock, an easygoing man with a sly sense of humor, that Prusiner's prion theory might be an example of what the Nobel laureate chemist Irving Langmuir once called "pathological science"; he suggested I look up a classic talk the General Electric pioneer gave on the subject back in 1953. The next day I climbed the hill to Yale's Klein Science Tower, blew the dust off an old volume of *Physics Today* and found the talk.

Langmuir, who died in 1957, made a hobby of collecting sensational scientific "discoveries" that turned out to be false—not fraudulent but merely self-deluded. Cold fusion, which came and went a decade ago, is probably the best-known recent example; if Langmuir had still been around, he would certainly have included it on his list. He mentions French physicist René-Prosper Blondlot's 1903 discovery of "N rays," which were supposed to be something like X rays but were given off by human bodies, bricks and other common objects; they had the odd effect of making it easier to see things in the dark. "They don't follow the ordinary laws of science that you ordinarily think of," Langmuir has Blondlot claiming. Once the French physicist reported his discovery, other scientists looked for N rays and found them as well. Unfortunately for Blondlot's reputation, a skeptic came along who covertly pocketed a crucial piece of apparatus, an aluminum prism, while the physicist was measuring N-ray intensities. Since the room was dark, Blondlot didn't realize he'd been sabotaged and went right on measuring what removing the prism guaranteed could no longer be there. The skeptic told the world. "And that," Langmuir concludes, "was the end of Blondlot."

Pathological science raises deep questions about human perception. "These are cases," Langmuir says charitably, "where there is no dishonesty involved but where people are tricked into false results by a lack of understanding about what human beings can do to themselves in the way of being led astray by subjective effects, wishful thinking or threshold interactions. . . . These are things that attracted a great deal of attention. Usually hundreds of papers have been published on them. Sometimes they have lasted for 15 or 20 years and then they gradually have died away."

Extracting what his cases have in common, Langmuir offers a list of "Symptoms of Pathological Science." I tried matching Langmuir's list to prion theory and found several obvious matches. Then Brock faxed me a paper he was drafting that identified several more.

"The effect is barely detectible" is one symptom. Since no one has demonstrated that prions alone are infective, Prusiner's theory has yet to cross this threshold—his effect isn't detectable at all. "The size of the effect is independent of the intensity of the cause" is a corollary symptom. PrP tracks with infectivity in crude samples of infective material, but when researchers try to purify a sample to concentrate the infectivity, infectivity goes one way and PrP level goes another.

"Claims of great accuracy" fits the prion's claimed ability to carry precise and various strain information. In fact, the prion, a middle-sized protein, is purported to be a one-man band, combining into itself the complex functions that a virus requires an orchestra of nucleic acid coated with proteins to perform: entering the body, avoiding bodily defenses, maneuvering to whatever organ it targets, invading a cell, hijacking the cell's genetic machinery and using it to pump out copies of itself.

That the prion alone in all of biology reproduces without nucleic acid is certainly a "fantastic theory contrary to experience." (To give Prusiner his due, he's always been careful to add, *sotto voce*, that some unidentified shadow substance might help it along. In one grant proposal which Gary Taubes uncovered, he even speculated that "a second component [might be] responsible for strain specificity, the obvious candidate being a nucleic acid." More recently, he has postulated an unidentified "Protein X" to fill this role.)

"Criticisms are met by *ad hoc* excuses thought up on the spur of the moment" certainly fits Prusiner's "second component" hedge, and there are many other examples. Langmuir's symptom "the ratio of supporters to critics rises up to somewhere near 50 percent" defines exactly where prion theory is today; whether that ratio "then falls gradually to oblivion" remains to be seen.

Infectious proteins, crystalline or prionic, may carry the day, or the TSE agent may turn out to be a virus. As I write, in November 1997, the evidence isn't in. But the ascendancy of the prion makes me nervous. It reminds me of how the Fore got themselves into trouble. They studied the new disease, kuru, that had appeared among them, as carefully as Carleton Gajdusek would study it later. Applying their culture's version of epidemiology, they concluded that kuru wasn't an ordinary sickness like diarrhea or pneumonia but a bewitchment, the consequence of sorcery. Sorcery was their prion, if you will, a colorful but anomalous cause. Having reached that conclusion (later revealed to be erroneous), they saw no reason to look further, no reason to take special precautions, no harm in eating the kuru dead. Thus they spread the disease among them to epidemic disaster.

To prevent a BSE epidemic in the United States, the U.S. Food and Drug Administration on June 5, 1997, published a final regulation that prohibited the use of most mammalian protein in the manufacture of animal feeds for cows, sheep and goats. The regulation took effect on August 5. Excluded from the ban were blood, blood products, gelatin, milk, milk products, pig and horse protein and inspected meat products offered cooked for human consumption, including restaurant and institutional plate waste.

That's a ban with exclusions big enough to drive a cortège of hearses through. The FDA's own TSE Advisory Committee, chaired by Paul Brown (Stanley Prusiner is a member), warned the agency in April 1997 that gelatin hadn't been ruled out as a TSE transmission channel and recommended stricter controls on gelatin imports from BSE countries. Brown reported at a World Health Organization conference in March that blood might also be in doubt. In laboratory experiments he was just completing he found that blood plasma from mice infected with CJD, if injected into the brains of healthy mice, transmitted the disease. WHO has recommended excluding from blood donation people at risk of CJD and other TSEs: anyone who has ever received native human growth hormone or gonadotrophin, anyone with a family history of TSE and anyone who has received a graft of human dura mater during neurosurgery. The Red Cross has complied. Blood accounts for about three billion of the forty-four billion pounds of slaughterhouse waste processed by U.S. rendering plants annually; the new FDA regulation allows it to be incorporated into animal feed.

"Sludge," the rendering industry's term for restaurant and institutional plate waste, accounts for another two billion pounds of rendering material in the United States annually. It includes, of course, beef, lamb and mutton scraps, spinal cord and bone. The FDA regulation allows it in animal feed as well.

British researchers have demonstrated pig susceptibility to BSE infection by direct inoculation into the brain. No pigs have yet turned up at market with unequivocal signs of TSE, but if such a disease were to occur in the species, its incubation period would probably be longer than the life span of pigs raised for commercial slaughter—which means no one would know if human transmission was occurring until humans began showing symptoms, years after the fact.

Rendered material processed into animal feed in the United States includes not only slaughterhouse waste, blood and sludge. It also includes cats and dogs euthanized by veterinarians and in animal shelters and pounds. It includes mink carcasses after the valuable fur is removed. It includes large animals such as deer killed on the road. These categories account for more than two billion pounds of rendered material annually. All these species are susceptible to TSE infection. Deer harbor a native TSE known as chronic wasting disease, and six percent of northern Colorado bucks killed by hunters in 1996 proved to be infected.

Farmers seeking to reduce costs have begun incorporating manure into their cattle feed, particularly farmers in California, the mid-Atlantic and the South with access to poultry manure from industrial chicken and turkey operations. Manure is rich in nitrogen, which animals use to make protein, and poultry-house litter loaded with manure costs five times less than good alfalfa hay. Farmers prepare the manured litter by composting it; fermentation heats it sufficiently to pasteurize it—but not, of course, to kill hardy TSE agents. In August 1997, *U.S. News & World Report* interviewed a cattle farmer near Dardanelle, Arkansas, who had recently purchased 745 tons of chicken-house litter rich with manure, had composted it for ten days and mixed it with a smaller portion of soybean bran, and was feeding it to his eight hundred head of cattle. "My cows are fat as butterballs," he told the newsmagazine. "If I didn't have chicken litter, I'd have to sell half my herd. Other feeds are too expensive." From infected cattle, then, or mink, or cats, or dogs, or deer, TSE agents could pass into protein supplement for poultry rations, thence into poultry manure, thence back into cattle as composted feed. Such scenarios are hypothetical, but they demonstrate that the new FDA regulations, offered as a "firewall" against transmission, will not certainly protect consumers if a BSE epidemic occurs.

TSE researchers generally agree that all species that make PrP are theoretically susceptible to TSE infection. Besides humans, cattle, mink, sheep, goats, cats, dogs, deer and elk, squirrels have now been implicated as carriers. Physicians have identified a cluster of eleven CJD victims in rural western Kentucky where only one case a decade is statistically probable. All the victims were eaters of squirrel brains, a regional delicacy. "Someone comes by the house with just the head of a squirrel," Dr. Erick Weisman told *The New York Times,* "and gives it to the matriarch of the family. She shaves the fur off the top of the head and fries the head whole. The skull is cracked open at the dinner table and the brains are sucked out."

These many familiar, suspected or newly revealed transmission channels demonstrate that spongiform encephalopathy is a risk to food animals not only in North America but everywhere in the world. It follows that the only certain way to avoid a food-borne TSE epidemic in humans—Collinge's "disaster of biblical proportions"—is to ban entirely the feeding of animal protein to animals.

Such recycling of slaughterhouse and other animal waste was developed during the Second World War to boost food production and eliminate a major source of environmental contamination: waste material, particularly blood, used to be piped out the back of slaughterhouses directly into rivers, one reason American rivers were polluted. Recycling animal protein back through animals was a successful technology (whatever you may think of its esthetics), as long as the rendering process killed all the pathogens in the material. Unfortunately, no rendering process reliably inactivates the TSE agent. The dilemma, then, is that banning all animal protein recycling would require finding some other way to dispose of some forty-four billion pounds of rotting animal tissue every year in the United States alone. Burning it in incinerators and burying it in landfills are the only means so far proposed.

Even eliminating rendering waste from the human food supply may not be sufficient to prevent a TSE epidemic. Disturbingly, the CJD Surveillance Unit in Edinburgh reported in 1997 the results of a survey of the eating habits of 187 people with sporadic CJD—the "random" kind that emerges in middle age uniformly throughout the world—identified in Great Britain between 1990 and 1996. The Edinburgh group found almost a threefold increase in risk of sporadic CJD associated with eating beef every month, and a 3.3-fold increase among those who eat beef weekly. People who ate brain even rarely had a fourfold higher risk.

Transmissible diseases, by definition, have to be transmitted. Some transmit through the air and many transmit through water. Air- and water-borne diseases have been relatively easier to control than sexually transmitted diseases such as syphilis and HIV. The TSEs, though their infectivity is obviously low, are the most insidious of all. Hidden in food, they're essentially unidentifiable until they progress to the point of brain damage, months or years after infection. Contamination-control measures can reduce the risk, but can no more eliminate it entirely than meat inspection alone can eliminate *E. coli* and *Salmonella*. The solution to bacterial contamination of food is irradiation, but irradiation doesn't work against the TSEs. If the TSE agent is a virus, then identifying the virus should make it possible to develop animal and human vaccines to protect us. That's why it's important to keep looking, and why Stanley Prusiner's doubtful Nobel Prize muddies the waters. Who's going to fund viral studies now?

Glade
November 1997

Readers interested in following TSE developments will find a valuable current record on the Internet at www.mad-cow.org. Thanks to Martin Brock, Paul Brown, Carleton Gajdusek and Laura Manuelidis for information and advice.

GLOSSARY

ALS, Lou Gehrig's disease: amyotrophic lateral sclerosis, a disease of the nervous system characterized by muscle irritability and wasting.

Alzheimer's disease: a noninfectious, progressive brain amyloidosis (*q.v.*) causing dementia and eventual death, named for the German neurologist Alois Alzheimer.

amyloid; amyloidosis (AM-ih-loyd; AM-ih-loy-DOE-sis): protein fibers that show characteristic green color in polarized light when stained with the dye Congo red; the disease caused by their accumulation in various organs and tissues, including the brain.

amyloid plaque: a microscopic tangled mass of amyloid (*q.v.*) in brain tissue.

amyloid (prion) rods: see SAF.

Anga (AHNG-a): The Papua New Guinea Eastern Highlands linguistic group formerly known as the Kukukuku (*q.v.*). Carleton Gajdusek coined the name, adapting the word common to several Anga dialects for "house."

antibody: a class of proteins in the body which react to invasions of foreign proteins such as microbes.

APP: amyloid precursor protein, the protein which forms amyloid plaques in Alzheimer's disease.

astrogliosis (ass-tro-glee-OH-sis): abnormal enlargement and proliferation of glia, the cells in the brain which fight infection.

athetoid movement; athetosis (ATH-ih-toid; ath-ih-TOE-sis): continuous slow involuntary movements of the hands and feet.

brain stem: the central trunk of the brain, which continues downward to form the spinal cord.

BSE (bovine spongiform encephalopathy) (en-seff-uh-LOP-uh-thee): a spongiform disease of cattle first identified in Britain in 1986.

cargo cult: the belief among various Western Pacific groups that the goods in American, Australian and other cargo ships and aircraft were stolen from the spirits of their ancestors, which would return someday and deliver the cargo to them.

cerebellum (sara-BELL-uhm): the "small brain" below and behind the larger body of the cerebrum (*q.v.*) which is responsible for balance, for muscle tone and for coordinating the timing of voluntary movements.

cerebral cortex: the outer gray matter of the brain.

cerebrum (suh-REEB-rum): the largest part of the brain, convoluted and divided like a walnut.

Compton: the British Agricultural Research Council Field Station at Compton in Berkshire, renamed the Institute for Research on Animal Diseases in 1963.

Creutzfeldt-Jakob (KROYTZ-felt YAHK-ohb) **disease (CJD):** a progressive, fatal degenerative disease of the brain characterized by massive incoordination, seizures and dementia. Spongiform damage to the brain and the presence of SAF (*q.v.*) are diagnostic of the condition, which is named for two German physicians who first published case reports.

degeneration: pathologic change causing impairment or destruction of function.

dementia: loss of mental function.

DNA: deoxyribonucleic acid, one of two nucleic acids that living organisms use as carriers of genetic information.

downer cattle: cattle which have fallen and are unable to stand because of disease.

EDT (ethylene diamine tartrate): a chemical compound used in industrial processing.

encephalitis (en-seff-uh-LIGHT-is): inflammation of the brain.

encephalopathy (en-seff-uh-LOP-uh-thee): disease of the brain.

endemic disease: disease habitually present.

endocannibalism: cannibalism of relatives.

epidemiology (ep-ih-deem-e-AHL-uh-gee): the study of disease outbreaks and epidemics.

fascia (FASH-uh): the fibrous coating of muscles.

fatal familial insomnia (FFI): a form of familial Creutzfeldt-Jakob disease (*q.v.*) which presents initially as persistent, morbid insomnia.

Fore (FOR-ae): a Papua New Guinea Eastern Highlands linguistic group, numbering about thirty thousand people, which suffered from kuru (*q.v.*); also, the geographic territory of the group.

formalin (FORM-ah-linn): a solution of thirty-seven percent formaldehyde stabilized with methanol.

Gajdusek (GUY-du-shek), **D. Carleton:** American Nobel laureate pediatrician and virologist.

gene: a unit of heredity controlling a single characteristic such as eye color.

genome (GEE-nome): the full complement of genes of an organism—some ten million in the case of humans.

glia (GLEE-a): a type of nerve cell that functions for insulation, infection-fighting and housekeeping.

Goroka (Go-ROH-ka): the capital of the Eastern Highlands Province of Papua New Guinea.

GSS (Gerstmann-Sträussler-Scheinker syndrome): a familial form of Creutzfeldt-Jakob disease. GSS progresses more slowly than typical CJD and produces multicentric amyloid plaques.

iatrogenic (ee-at-tro-GEN-ik): literally "physician-born": unintentionally induced by physicians.

in vitro (in VEE-tro): "in glass"; said of biological experiments conducted outside an organism; test-tube experiments.

infantile paralysis: polio.

intention tremor: tremor induced by voluntary movement of a limb.

Kainantu (Kai-ih-NAN-too): a town in the Eastern Highlands of Papua New Guinea.

Kukukuku (koo-KOO-kuh-koo): old, pejorative name for the Anga of the Papua New Guinea Eastern Highlands.

kuru (KOO-roo): a progressive, fatal brain disease, spread by cannibalism, confined to the Papua New Guinea Eastern Highlands and there almost exclusively to the Fore people (*q.v.*). The word means "trembling" and "fear" in Fore.

kuru plaque: a large, chrysanthemum-like amyloid plaque (*q.v.*) characteristic of kuru and of new-variant CJD.

louping ill: a disease of sheep which causes them to spring up and down when moving forward.

lumbar puncture: a tap of the spine in the region of the lower back to sample spinal fluid.

MAFF: the British Ministry of Agriculture, Fisheries and Food.

multiple sclerosis (sklare-RHO-sis): a chronic degenerative disease of the central nervous system which destroys the insulating sheath of nerves, causing muscle weakness, loss of coordination and speech and visual defects.

neuropathology: the study of diseases of the nervous system, particularly by examining the tissue itself.

new-variant CJD: a new form of Creutzfeldt-Jakob disease believed to be transmitted by eating the tissue of cattle infected with BSE (*q.v.*); vCJD.

Okapa (oh-KAHP-ah): a patrol post and administrative center in the middle of the Fore region of the Papua New Guinea Eastern Highlands.

Parkinson's disease, Parkinsonism: a disease characterized by tremor and physical rigidity caused by damage to the substantia nigra area of the brain, where dopamine is produced.

prion (PREE-on): a term proposed by neurologist Stanley Prusiner to identify "small *pro*teinaceous *in*fectious particles which are resistant to inactivation by most procedures that modify nucleic acids." The term, Prusiner added, "underscores the requirement of a protein for infection; current knowledge does not allow exclusion of a small nucleic acid within the interior of the particle."

PrP: prion (*q.v.*) protein or protease-resistant protein.

reagent (re-AGE-ent): any substance used in chemical reactions, particularly to test for the presence of another substance.

Rickettsia (rick-ETT-see-ah): rod-shaped parasitic bacteria, a genus named for the American pathologist H. T. Ricketts, who first identified them. Typhus is a rickettsial infection.

RNA: ribonucleic acid, one of two nucleic acids that living organisms use as carriers of genetic information.

SAF (scrapie-associated fibrils): stacks of PrP (*q.v.*) crystals visible under the electron microscope as twisted fibers in homogenates of brain tissue infected with transmissible spongiform encephalopathy.

scrapie (SCRAPE-ee): a natural progressive brain disease of sheep that frequently causes itching so intense that the animals scrape off their wool seeking relief.

Smadel (smay-DELL), **Joseph:** American virologist and administrator; associate director of the U.S. National Institutes of Health.

spongiform (SPONGE-ih-form): resembling a sponge in structure; full of holes.

status spongiosus (spun-gee-O-sus): extensive spongiform (*q.v.*) damage.

status epilepticus (ep-ih-LEP-tee-cuss): rapidly recurring epileptic seizures.

strain: a variant form of a disease such as CJD that results from mutation of the disease-agent genome.

TME (transmissible mink encephalopathy): a feed-borne spongiform disease of mink.

transmissible spongiform encephalopathy (TSE): generic name for the various spongiform diseases of animals and humans, including kuru, Creutzfeldt-Jakob disease, scrapie and TME, all characterized by spongiform pathology and the presence of SAF (*q.v.*).

virino (vy-REEN-o): a hypothetical viral disease agent consisting of a short piece of DNA which recruits host machinery to manufacture its protein coat and to reproduce.

virology (veer-OL-o-gee): the scientific study of viruses.

Wanitabe (wan-ih-TAHB-uh): a village in the South Fore region of the Papua New Guinea Eastern Highlands where anthropologists Robert Glasse and Shirley Lindenbaum lived during a South Fore field study.

ACKNOWLEDGMENTS

Carleton Gajdusek generously shared his papers and journals and gave me days of interviews at a time of great personal turmoil.

For interviews, documents and information, thanks to Joe Gibbs and Paul Brown at the NIH; Bill Hadlow, Bruce Chesebro and Byron Caughey in Hamilton, Montana; Shirley Lindenbaum in Manhattan; Patricia and George Merz on Staten Island; Alan Dickinson, Hugh and Janet Fraser in Edinburgh; John Collinge in London; Richard Lacey in Leeds; Mike Alpers in Sydney; Fred Brown on Plum Island; Raymond Hintz in Palo Alto; John Honstead at the FDA. Charles Mgone welcomed me to Goroka and found time to arrange my trip to the South Fore. George Koki drove the hard four-hour drive and guided my Waisa tour. Campbell Crilly in Cairns patched up Ginger. Gil and Anita Elliot prepared the way and made my London visit a pleasure. Miranda McMinn shared documents. Malcolm Withers and Kenteas Brine arranged a backstairs view of Parliament.

Thanks to Doron Weber and the officers of the Alfred P. Sloan Foundation for providing and to Mary Freedman and Larry Norton of the Pequot Library for administering a timely travel grant.

George Klein in Stockholm has been generous with advice and insight and reviewed the book in manuscript. So did my daughter, Kate, and her husband, Jacques Perrault, both molecular biologists. Norman Charles advised on ophthalmology. Any errors that survive their scrutiny are my own. Gail Harris at Yale's Cushing Medical Library tracked down hundreds of papers. Helen Haversat skillfully transcribed interviews dense with neurological and biological jargon.

Mort Janklow and Anne Sibbald lit the torch and kept it supplied with fuel. Michael Korda guided the work with his usual finesse.

My wife, Ginger Rhodes, organized our travel, recorded the interviews, photographed, supervised transcription and reviewed every chapter. She deserves the dedication anyway, but she earned it.

INDEX

Page numbers in *italics* refer to maps and photographs.